Ronald Müller

Algorithms and System Aspects for Automatic Face Analysis

Ronald Müller

Algorithms and System Aspects for Automatic Face Analysis

Performance and Challenges of Modern Face Analysis Techniques Built on Current Standard Hardware and Latest Mathematical Algorithms

Südwestdeutscher Verlag für Hochschulschriften

Impressum/Imprint (nur für Deutschland/ only for Germany)
Bibliografische Information der Deutschen Nationalbibliothek: Die Deutsche Nationalbibliothek verzeichnet diese Publikation in der Deutschen Nationalbibliografie; detaillierte bibliografische Daten sind im Internet über http://dnb.d-nb.de abrufbar.
Alle in diesem Buch genannten Marken und Produktnamen unterliegen warenzeichen-, marken- oder patentrechtlichem Schutz bzw. sind Warenzeichen oder eingetragene Warenzeichen der jeweiligen Inhaber. Die Wiedergabe von Marken, Produktnamen, Gebrauchsnamen, Handelsnamen, Warenbezeichnungen u.s.w. in diesem Werk berechtigt auch ohne besondere Kennzeichnung nicht zu der Annahme, dass solche Namen im Sinne der Warenzeichen- und Markenschutzgesetzgebung als frei zu betrachten wären und daher von jedermann benutzt werden dürften.

Verlag: Südwestdeutscher Verlag für Hochschulschriften Aktiengesellschaft & Co. KG
Dudweiler Landstr. 99, 66123 Saarbrücken, Deutschland
Telefon +49 681 37 20 271-1, Telefax +49 681 37 20 271-0, Email: info@svh-verlag.de
Zugl.: München, Technische Universität München, Dissertation, 2008

Herstellung in Deutschland:
Schaltungsdienst Lange o.H.G., Berlin
Books on Demand GmbH, Norderstedt
Reha GmbH, Saarbrücken
Amazon Distribution GmbH, Leipzig
ISBN: 978-3-8381-0366-2

Imprint (only for USA, GB)
Bibliographic information published by the Deutsche Nationalbibliothek: The Deutsche Nationalbibliothek lists this publication in the Deutsche Nationalbibliografie; detailed bibliographic data are available in the Internet at http://dnb.d-nb.de.
Any brand names and product names mentioned in this book are subject to trademark, brand or patent protection and are trademarks or registered trademarks of their respective holders. The use of brand names, product names, common names, trade names, product descriptions etc. even without a particular marking in this works is in no way to be construed to mean that such names may be regarded as unrestricted in respect of trademark and brand protection legislation and could thus be used by anyone.

Publisher:
Südwestdeutscher Verlag für Hochschulschriften Aktiengesellschaft & Co. KG
Dudweiler Landstr. 99, 66123 Saarbrücken, Germany
Phone +49 681 37 20 271-1, Fax +49 681 37 20 271-0, Email: info@svh-verlag.de

Copyright © 2009 by the author and Südwestdeutscher Verlag für Hochschulschriften Aktiengesellschaft & Co. KG and licensors
All rights reserved. Saarbrücken 2009

Printed in the U.S.A.
Printed in the U.K. by (see last page)
ISBN: 978-3-8381-0366-2

Contents

Contents		1
1 Introduction		**1**
1.1	The Challenge	1
1.2	*FEASy* – a FacE Analysis System	2
1.3	The Book in a Nutshell	4
2 A Multi-Threading Framework for Signal Processing Systems		**5**
2.1	Conditions for the development of Signal Processing Systems	6
2.2	Other works	7
2.3	Requirements of a Software Framework for High-Performance Signal Processing Systems	8
2.4	Concepts of MMER_Lab	9
	2.4.1 Software Architecture	12
	2.4.2 Design Decisions	13
2.5	Application Examples and Evaluation	13
2.6	Conclusion	15
3 Object Localization with AdaBoost Variants on Haar- and Gabor-Wavelet Features		**17**
3.1	Haar-like and Gabor-Wavelet features	18
	3.1.1 Haar-like features	18
	3.1.2 Gabor-Wavelets	18
3.2	Feature Selection and Classification with AdaBoost	22
	3.2.1 Notation	22
	3.2.2 The Standard AdaBoost Algorithm	23
	3.2.3 Gentle AdaBoost	24
	3.2.4 Weak classifiers	25
	3.2.5 Cascaded AdaBoost Classification	27
3.3	Evaluation of Localization Performance	28
	3.3.1 Databases	29
	3.3.2 Head and Eye Localization Results	29
	3.3.3 Feature selection	30
	3.3.4 Localization Performance	31
3.4	Conclusion	32

Contents

4 The Theory of Active Appearance Models — 35
- 4.1 Preparation of Training Data — 36
  - 4.1.1 Alignment and Normalization of Landmarks — 36
  - 4.1.2 Warping — 38
  - 4.1.3 Normalization of Textures — 39
- 4.2 Generation of an Appearance Model — 41
  - 4.2.1 Shape Model — 41
  - 4.2.2 Texture Model — 42
  - 4.2.3 Combined Model — 42
- 4.3 Coefficient Optimization — 43
  - 4.3.1 Objective Function — 44
  - 4.3.2 Offline Prediction — 46
  - 4.3.3 Numerical Estimation of the Jacobian Matrix — 47
  - 4.3.4 Iterative Optimization — 49

5 Derivatives and Advancements of Active Appearance Models — 51
- 5.1 A Survey on Active Appearance Models and Variants — 51
- 5.2 Appearance Models based on NMF — 53
  - 5.2.1 Data Modeling with Non-Negative Matrix Factorization — 54
  - 5.2.2 Generation of Appearance Models with NMF — 66
  - 5.2.3 Conclusion — 69
- 5.3 Online Optimization of AAM Coefficients — 70
  - 5.3.1 Gradient Descent — 71
  - 5.3.2 Grid Sampling — 73
  - 5.3.3 Nelder-Mead or Simplex Optimization — 74
  - 5.3.4 Conclusion — 76
- 5.4 GPU-Accelerated Active Appearance Models — 77
  - 5.4.1 Warping — 77
  - 5.4.2 Coefficient Optimization — 83
  - 5.4.3 Conclusion — 86
- 5.5 Evaluation Measures for the Quality of AAM Re-synthesis — 86
  - 5.5.1 Dataset Annotation — 87
  - 5.5.2 Quality Measures — 88
  - 5.5.3 Evaluation of Quality Measures — 90
- 5.6 Summary — 91

6 Application of Active Appearance Models to Face Analysis — 93
- 6.1 Classification Based on Results of the AAM Optimization — 94
  - 6.1.1 Classification based on class specific AAMs — 94
  - 6.1.2 Statistical classification based on AAM coefficients — 95
  - 6.1.3 Support Vector Machines — 96
  - 6.1.4 N-fold Cross-Validation — 97
- 6.2 Image Databases — 98
  - 6.2.1 The AR Database — 98
  - 6.2.2 The NIFace1 Database — 99

	6.2.3	The FG-NET Aging Database	99
	6.2.4	The MMI Face Database	100
6.3	Gender Recognition		101
	6.3.1	Dataset	101
	6.3.2	Results	101
	6.3.3	State-of-the-Art	103
6.4	Facial Expression Recognition		104
	6.4.1	Dataset	104
	6.4.2	Results	104
	6.4.3	State-of-the-Art	106
6.5	Person identification		107
	6.5.1	Dataset	107
	6.5.2	Results	107
	6.5.3	State-of-the-Art	108
6.6	Head Pose Recognition		108
	6.6.1	Dataset	108
	6.6.2	Results	109
	6.6.3	State-of-the-Art	116
6.7	Age Recognition		117
	6.7.1	Dataset	117
	6.7.2	Results	118
	6.7.3	State-of-the-Art	120
6.8	Comparison with NMF-AAMs		120
6.9	Conclusion		121

7 Summary — 127

I Appendix — 129

A Conventions — 131

A.1	General Typesetting		131
	A.1.1	Indexing	131
	A.1.2	Sets	131
	A.1.3	Scalars	131
	A.1.4	Sequences	131
	A.1.5	Functions	132
	A.1.6	Transformations	132
	A.1.7	Vectors	132
	A.1.8	Matrices	133
	A.1.9	Code	133
	A.1.10	Text Substitution	133
A.2	Symbols		134

List of Figures — 140

List of Tables	142
List of Listings	143
Bibliography	145

Chapter 1

Introduction

1.1 The Challenge

In human-human interaction the visual communication channel contributes immensely to the efficiency and quality of the communication [75], since the nuances are indeed transported via the tune of the voice but primarily by the facial expression, the body pose, head pose, and gestures. For a good understanding the *context* is essential. As O. Wiio humorously states in his *laws* ("Communication usually fails, except by accident"), a working communication requires the partners to ideally have the same context or at least know the context of their opponent. The exchange of pure verbal content regularly leads to mis-interpretations and mis-understandings. Thus, for modern and future technical applications which strive to analyze human-human communication and improve human-computer interaction, the analysis of faces is mandatory where already basic difference in gender and age can give identical words a totally different meaning. Finally a human face itself provides various information about the person: gender, age, ethnic origin, ametropia, and identity. The overall target of this work is the extraction of as much information as possible from a face at high computational speed.

However, the task of face analysis with whatever scope is characterized by highly complicated conditions. In the first place, the natural appearance of human faces varies in their outer shape, the relative spatial position of facial features (eyes, brows, nose, mouth, chin), the shape of the facial features, and the skin color and texture.

Due to the high complexity of face recognition and interpretation, the evolution even dedicated an own area of the human visual brain cortex to this task [58].

Therefore, human face analysis constitutes one of the greatest challenges in the modern computer vision area. Our face analysis technology utilizes the basic idea of the Active Appearance Model [26] approach: A statistical model describes the variations in shape and texture of human faces derived from a careful selection of photographs showing different persons with different facial expression and in various lighting conditions. During the analysis of a human face within a video or a picture, the Appearance Model is used to re-synthesize this face as optimal as possible. Finally, this technology provides an extremely compact parametrization of the analyzed face by setting between 20 and 60 numeric parameters. This delivers a highly compact encoding of the face properties which can be decoded by subsequent statistical classification methods.

1.2 *FEASy* – a FacE Analysis System

During the research for this book a software application eventually named the FacE Analysis System *FEASy* emerged. Face analysis is a just as complex as valuable component of systems for human behavior analysis or for improving the attractiveness and usability of Human-Machine-Interfaces. Thus, a considerable amount of manpower was invested in the implementation of a usable, widely configurable, extendable, and high-performance software system in order to allow for our research findings having an effect in larger systems addressing a higher semantic level. Since the entire system was developed by a team of programmers and currently provides the selection of 12 algorithmic variants with permutations and the adjustment of 76 relevant algorithmic parameters, not to mention the compatibility with a bunch of input data options, a well structured software architecture with several abstraction layers, object-orientation, and modularization is mandatory [33, 45].

Therefore, *FEASy* is built as modular processing chain in the multi-threading framework MMER_Lab (see chapter 2) and consists of the following components:

- Image Source:
  Image-Files, video files, and cameras can serve as input. Multiple image formats (JPEG, TIFF, BITMAP, Portable formats), video codecs (several Microsoft and Intel codecs, DivX, Cinepak, etc.), and all cameras with DirectX drivers for MS Windows or Linux drivers are supported [45]. In case of camera or video file sources, each frame is provided as single image to the subsequent processing chain.

- Head- and Eye-Tracking:
  Presuming that there is a face with both eyes visible to the camera in each image, this module localizes the position of the face in a first step. Thereby an algorithm is used which applies an AdaBoost feature selection and classification approach on Haar-Wavelet features [115]. This method is known for its low computational complexity, robustness against illumination variations, and its applicability to all ethnic groups with a single model. However, it shows weaknesses when it comes to horizontal head rotation or rotations in the image plane. Relying on the head localization, a localization of both eyes is performed, again applying an AdaBoost algorithm. For the purpose of a precise eye localization we add Gabor-Wavelet features which are more computationally intensive but showed higher localization accuracy. Details about the research, the applied algorithms, and evaluations are given in chapter 3. The output ports of this module provide the absolute pixel

coordinates of the head and both eyes which is required as initialization of
the subsequent face re-synthesis.

- Face Re-synthesis:
  This module constitutes the core of the system in several ways. Initially, it
  generates over 90% of the computational costs. Further, the topic of this
  work evinces that also from the algorithmic point of view this is the crucial
  and most challenging part of the system where most research efforts were
  invested. The chapters 4 and 5 address the starting basis and enhancements
  of the applied algorithms in statistical shape and texture models while chapter 6 shows possible applications with evaluation results. The implemented
  approach relies on the generation of statistical models of the appearance,
  i.e. the shape and texture, of a face. Such Appearance Models allow the
  synthesis of a variety of instances of faces by adjusting a set of scalar coefficients. The analysis is performed by a re-synthesis of an unknown face
  via multivariate optimization of the model coefficients.

  The optimized coefficient values constitute a low-dimensional representation
  of the analyzed face for further processing and are provided as output of
  the module.

- Classification modules:
  The low dimensional representation of the face to be analyzed serves as
  feature vector for the again *statistical* recognition of facial properties. In
  context of this work we target on abstract information about the gender,
  age, identity, head pose, or facial expression of a person. In accordance
  to research works in the area of statistical classification we could confirm
  that Support Vector Machines (*abbrev.* 'SVM') provide the best recognition performance at comparably low computational complexity during the
  recognition and with limitations of sparse data [101]. For each desired recognition task, a classification module must be dedicated with the according
  SVM models.

- Display:
  For demonstration purposes e.g. of the shape and texture generated by the
  face re-synthesis module, the avatarization of the analyzed face with foreign
  face textures, or of the classification results, this module can visualize the
  input and results of *FEASy*.

- Communication:
  If the system is run as a component e.g. of a multi-modal system where
  possibly speech or sensor data provide additional information about a person [38], this module broadcasts the results of the visual analysis via socket
  connections.

The configuration of *FEASy* can be conducted at runtime via the Graphical
User Interface of MMER_Lab, or a-priori by loading a correspondingly edited

XML configuration file. Apart from the remarkable accuracy in face analysis (see chpater 6), our system features high computational efficiency due to the application of hardware optimized mathematic libraries and the involvement of the latest graphic card platforms at the adequate steps in the algorithm (refer to section 5.4). While *FEASy* constitutes the basis of our evaluations of the developed face analysis algorithms, it proved its value as impressive demonstration application at several occasions [80, 81].

1.3 The Book in a Nutshell

The structure of this book follows the layout of the FacE Analysis System described above. Since the framework MMER_Lab constitutes the basis of the implementation and allows for the flexibility and computational performance of *FEASy* it is introduced at the beginning in chapter 2. Since automatic face analysis naturally requires a preceding localization of the face, this is presented in chapter 3. Chapter 4 describes the theoretic starting basis of our developments, the *Active Appearance Model* approach while chapter 5 presents a selection of algorithmic enhancements to the basic algorithm. Finally, a comprehensive, comparative evaluation on various databases is given in chapter 6 followed by a summary of the book and an outlook. The Appendix provides an overview on the typesetting, mathematical notation, and the symbol table.

Chapter 2

A Multi-Threading Framework for Signal Processing Systems

The target software application of this work on real-time face analysis constitutes a representative example of a digital signal processing system. It is serially assembled from various processing technologies, shows highest computational complexity on large data streams, and shall blend in larger software applications, at an adequate state of maturity. These characteristics demand for comprehensive solutions for the distributed software development process in teams, the system design sensitive to hardware exploitation and memory consumption, as well as for the experiencability and reliability of the implemented systems.

With MMER_Lab a software framework is available, which addresses a simplified system development process, re-usability of code, exploitation of modern multi-core CPU architectures, and user friendliness, all under utmost preservation of flexibility and computational performance. This is mainly achieved by a software architecture with multi-layered abstraction, the encapsulation of a multi-threading environment behind a clear Application Programming Interface (*abbrev.* 'API'), a system scheme of modules and synchronized data cables, and a scriptable as well as a graphical user interface for all levels of system developers to pure users. MMER_Lab proved its capabilities as basis of multiple systems of the Institute for Human-Machine Communication of the Technische Universität München.

The following section contemplates the current conditions in the development process of software based Signal Processing Systems (sec. 2.1). Section 2.2 introduces several attempts to support this process. The mentioned conditions and the specific properties of our face analysis system originate a set of requirements (sec. 2.3), which guided the concepts (sec. 2.4) and implementation of MMER_Lab. Section 2.5 presents different application scenarios, system setups, and an evaluative comparison to a competitive software framework. Finally, this invokes the conclusions of section 2.6.

2.1 Conditions for the development of Signal Processing Systems

The development process of large software based Signal Processing Systems (*abbrev.* 'SPS') is characterized by the following general properties.

Most SPSs can be described by flow charts of a data stream which is provided by a source, processed by several different filter modules, and finally disembogues into a sink. Thereby, multiple data streams may derive from different sources and are fused or flow into several sinks correspondingly. The source, filter, and sink modules are predominantly serial connected, although they may contain feedback loops in certain cases.

Furthermore, especially systems for audio, visual, or multi-modal signal processing, require expert knowledge from numerous different areas. This factor as well as the considerable programming effort for the build-up of large SPSs demands for the collaborative software development of researchers in teams. Consequently, the management of the development process with software engineering techniques, such as architecture concepts, interface definitions, testing, and maintenance gets mandatory or even necessary.

While SPS engineering requires a deep insight in the comprised technical areas, the integration of existing libraries from the world-wide community of programmers saves own man-power substantially, instead of re-inventing the wheel over again.

When it comes to latest signal processing methods in particular on speech-, image-, and video-data, immense amounts of data have to be handled leading to the corresponding computational effort. While, the hardware industry has been fighting the 4 GHz frontier of CPU clock speeds for years, the common approach points toward parallel processing with currently dual-, in near future multi-core CPU architectures. Unfortunately, standard implementations of serial SPSs can hardly benefit from this tendency since there is still no perceivable success in parallelized execution of source code by adequate compilers. On the other hand, parallelization can be realized via internal parallelization of algorithms, predominantly by separating logically independent functions such as large matrix operations, which often comes with an expensive and difficult reworking of the algorithms.

The main conditions and problems of the development of software based Signal Processing Systems are summarized under the following bullets:

- Size, technical diversity, and complexity demand for team collaboration and software engineering.

- The integration of a large amount of existing libraries makes re-implementations obsolete.

- The benefit from multi-core hardware can only be realized via parallel execution of the modules within mostly serial processing chains.

2.2 Other works

Several academic and industrial R&D teams have published systems, which try to provide support and solutions for a well-organized and effective software development process under the conditions mentioned in section 2.1.

The probably most established and spread framework is MATLAB (R) from MathWorks Inc. featuring a simple programming interface and numerous extension packages for various kinds of signal processing, simulation, and evaluation. This rapid prototyping software product is easy to learn and especially valuable for feasibility tests. The flexibility of the programming interface however comes with an remarkable overhead of mid-level code during execution time. Together with a sub-optimal memory management of MATLAB(R)'s Java back-end, the runtime performance of algorithms are way from native, even non-optimized C/C++ implementations. Beneficially, MATLAB(R) code is cross-platform compatible, which is important to be accepted by all, Windows(R), Linux, and MacOS users.

Another upcoming but cost intensive application is LabVIEW(R) from National Instruments Corporation. This commercial framework comes with a great variety of modules for a high range of signal processing tasks and the support of diverse external sensors or devices. It's interface for pure "graphical programming" allows for a system design without the barriers of code programming for all common operating systems. Although a large amount of functions and signal processing algorithms are already available, the addition and extension of modules requires a deep understanding of the flexible, though complex API of LabVIEW(R). The latest version 7.1 of LabVIEW(R) supports threading techniques for hyper-threaded and multi-core CPU hardware with an incorporated queuing system for the thread communication.

The most elaborated non-commercial framework for exploitation of PC clusters in high-end networks was developed by the National Institute for Standards and Technology (NIST), and is published as Smartflow 2 (*abbrev.* 'SF2') [77]. It is designed to distribute computationally intensive SPSs on a set of computers and uses socket communication for the data exchange of system components. Furthermore, SF2 also implements a multi-threading concept when several modules are started at a single PC. This concept is derived from MMER_Lab and was successfully integrated in SF2 [33] by one of the MMER_Lab authors Lukas Diduch [31]. One of the principles of SF2 is the decoupling of data types from data processing and flow. Thus, the framework is applicable to any kind of data, although it was meant for audio-visual processing.

Apart from these commercial systems several other parties build up their own solutions. ParleVision [95] is a framework for computer vision programming on Windows only. This framework allows to build processing chains, based on filters which support modularity. It has a very good performance in sequential processing but does not support parallelization in any way.

The two other systems, Fermus [74] and Smartflow [77] for network distributed computing applications are process-based utilizing IPC techniques as socket communication and shared memory. These systems, designed to work on networks, support modularity and the usage of multiple processors very well. However, context switching on standalone computers using multi-tasking decreases the performance compared to multi-threaded architectures. This is because multi-tasking systems operate within their own virtual address space. Processes are protected by the operating system from interference by other processes. A user process can not communicate with another process unless it makes use of IPC mechanisms.

Passing the data between computers also comes with network latencies which can affect real time high bandwidth computations such as video processing. The main advantage of these systems is the enormous computing power of a cluster versus the network latency and intercommunication time.

2.3 Requirements of a Software Framework for High-Performance Signal Processing Systems

The multi-threaded MMER_Lab was designed as a flexible system that can embed, combine and demonstrate computing intense algorithms. Mainly algorithms from the field of pattern recognition and signal processing (audio and video) were considered. It is of course also possible to embed algorithms from other scientific disciplines, but the base set of modules, which is distributed with the framework, was emphasized on the first field.

In the first place, the framework is used to provide a system for emotion recognition based on visual and statistical data utilizing AAMs and a classifier based on SVMs. Since the visual data emerging from a single source is less significant than data derived from multiple sources (e.g., additional audio sources), the system should be embeddable into a multi-modal and multiuser context for future research tasks.

To meet these requirements, the main idea was to develop a modular signal processing framework, providing sources, filters and sinks, and to embed this system into a multi-threaded environment. These sources, sinks and filters are represented as modules, which can be combined with each other. Each module has a computation core and a fixed interface. Multi-threading provides a natural way in expressing asynchronous signal processing systems. This helps in modeling real world problems.

Besides this core application, several side requirements also needed to be met:

- **Multi-platform capabilities**, which means that UNIX and Windows platforms can be used. This is an important feature, because it allows using the framework on a large number of computers.

- The ability to **embed every kind of software library** written in C or C++ into the system architecture to provide flexibility and adaptation in coding. As an example, high performance libraries for mathematical computation can be used instead of slow standard solutions.

- A **graphical user interface** (GUI) for control, as well as for editing and demonstration tasks. The GUI should be intuitively designed, to allow people to work with it after a short amount of time.

- The possibility to **automatize tasks** via scripting. This would allow any user to access the whole functionality of the framework without the need to actually control it in real time. Another requirement was that framework setups (i.e., networks of modules) should be able to be generated by scripts.

- The possibility to **log and monitor** results. In a multi-threaded environment, where various algorithms run in parallel, it is very important to keep track of what happens when, and to save the final results for later use or evaluation.

2.4 Concepts of MMER_Lab

'If you have an application that can benefit from parallel processing, make threading a priority in your company. The benefits can be enormous.[...]Plan for multi-core environments - not just dual-core. Assume there will be 4, 8, and even 16 cores in the future.'

Inspired by this citation from *Intel(R) Software Insight* magazine, issue on *Multi-core capability* [55](p.7), we started to conceptualize and implement the software framework MMER_Lab (lab.mmer-system.eu) in order to benefit from the new computational performance of the upcoming multi-core CPU and multi-CPU architectures in modern and future computer systems. In general this benefit can only be exploited via internal parallelization of algorithms mainly by separating logically independent functions such as large matrix operations. This often comes with an expensive and difficult re-implementation of the algorithms. For many systems with certain properties, however, there is an opportunity to profit from parallel computational power without re-implementation. In particular, Signal Processing Systems with several, mostly serial-connected components show a great potential for acceleration via porting into MMER_Lab framework. (Please note: We distinguish between *framework* and the *system* which runs in the framework.)

Frame-based video processing systems constitute an ideal example: Imagine a simple video analysis system consisting of four modules: a video source, a head localization and tracking module, a face-based person recognition module and a video display for demonstration. Such system can be parallelized by executing each module in its own thread. Thus, the person recognition module can perform its recognition task for frame number n on CPU-core 1, while the head tracking module already localizes the head(s) in frame m with $m > n$ using CPU-core 2.

However, threading requires considerable skills in programming. MMER_Lab is designed to be of benefit for all levels of software developers up to pure users. Thereby, module developers are not bothered by the problems and barriers of multi-threading implementation, since the embedding of algorithms is performed by usage of MMER_Lab's interfaces. Thanks to the Graphical User Interface (GUI), casual users can instantly load implemented modules, connect them to a complete system, and execute it via mouse clicks for ad-hoc demonstration or education purposes. Advanced users benefit from the scripting facilities and can perform numerous iterations and evaluations, e.g. of system parameter settings, in a batch processing style.

Our framework is developed cross platform for Linux and Windows operating systems and independent of the bit- architecture (32 or 64). This is achieved by purely gcc-compliant C/C++ programming code and usage of the Tcl/Tk [88] script interpreter. Furthermore, every kind of software library written in C and C++ can be applied for the implementation of modules for any signal processing system.

Thus, our framework provides an efficient opportunity to quickly transfer existing signal processing systems into multi-threaded applications with a full exploitation of the parallel clock speed of current and future multi-core architectures.

In the subsequent sections we explain and discuss the concepts of MMER_Lab for multi-threading, module structures and data flow, followed by application scenarios. Moreover we discuss our design decisions for MMER_Lab and give a conclusion.

The framework structure is based on the following three main concepts: A multi-threaded environment, a Tcl script interpreter with a Tk GUI, plus the underlying flexible and modular software architecture.

In the **Multi-Threaded Environment** (figure 2.1) modules are executed quasi parallel, depending on the number of available processors. Herein, the system modules embed algorithms and work like stand-alone processors with a fixed interface. Each module core runs as a thread with an additional thread for the modules communication interface. These are connected with cables (queues) for the transport of content (data, memory pointers, parameters, or commands). The state of both, cables and modules is protected by *mutexes* (short for *mutual exclusion*)[50]. Furthermore, *conditions* assume the synchronization of both object classes. This does not only avoid useless polling (a mechanism for status requests of a resource) from within the threaded functions, but also provides a powerful technique to synchronize the data flow and split the execution onto multiple processors. Sources push data into a cable, while filters and sinks pull the data from a cable. A source module stops *automatically* to produce content, if the outgoing cable buffer is full, and continues when the cable buffer is depleted. A sink module *automatically* stops to pull content from the incoming cable buffer if it is empty and continues as long as the cable buffer is filled with content. A filter module is a combination of both behaviors. The main synchronization part is implemented inside the cable and module classes using the *monitor pattern*

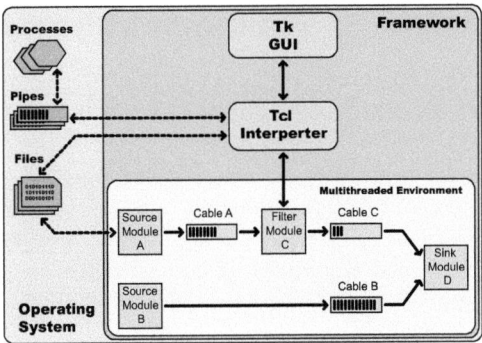

Figure 2.1: Schematic view of the framework concept

[100]. Multiplexers inside the modules interface are used for data dispatching with *round robin* or *priority based* methods.

The second important concept is the embedded **Tcl interpreter** including the **Tk GUI** (Fig. 2.1). The interpreter has been extended by new objects, like module and cable objects and a set of Tcl procedures to manipulate them. Firstly, this provides a user interface for communication with the multi-threaded environment. Note that the Tcl interpreter is the only facility of the MMER Lab which runs as the main procedure in the executed framework process. The interpreter itself is not multi-threaded or uses threading in any way. This is mentioned here, because Tcl can be compiled with threading support as well which would allow the execution of multiple Tcl interpreters in their own thread. We do not use this feature in order to maintain performance. Secondly, by using the Tcl scripting language [88], we can also create a *high level framework language*: A large set of control command wrappers, helper functions and Tk procedures for GUI control have been implemented this way.

This approach has several advantages: The Tcl/Tk language can be used to natively perform interface handling, common scripting tasks, and computations. Examples are file handling, creation of pipes to processes, computing numeric parameters for the system/module setup, and managing socket ports for distributed cluster processing. The script interpreter is also extremely helpful to configure module/cable setups for simplified loading and starting frequently used systems. No recompilation or restart is needed for setup modifications or completely different setups. During demonstration, practical courses and lectures this opportunity turned out to be very valuable. Additionally, a great variety of freely available Tcl/Tk packages can be used by the framework. Exemplarily, we load a plot-package for automatic generation of signal course or evaluation diagrams.

One important feature is the efficiency of plugging module packages into the system at *runtime*. Thus, memory and performance can be saved, because the module code (C/C++) is stored in a shared library which will only be loaded into memory on demand, including the dependent shared libraries. MMER Lab is per se started with no functionality except for instantiating and controlling cables and modules.

As mentioned above the front-end to the framework is implemented as a Tk GUI. However a GUI is *not required* to run the framework. It can be run in a mode only using scripts for setup and control. This feature is useful to gain utmost performance or hide the framework interface if desired.

The last core concept is a flexible **Software Architecture** which is explained in detail within the following section. Based on this architecture, the software developer is able to embed every existing library written in C/C++ into the modules' API. Thus, useless and laborious re-implementation of standard functionality becomes dispensable.

2.4.1 Software Architecture

The underlying software architecture of MMER_Lab consequently follows the concepts of procedural and data abstraction [4]. Four fixed layers of abstraction have been established to provide access from low level functions (libraries) up to high level procedures (modules).

The lowest system layer (**libraries**) contains all basic *external* code libraries used in the framework and in most of our systems. This includes standard libraries like the C++ stdlib, Xlib, OpenCV, OpenGL and Boost [60] or ATLAS, C-BLAS and GSL, high performance and high precision mathematical libraries compiled optimized for the dedicated processor architecture in Fortran, assembler and C.

The mid level system layer (**core functions**) is used to implement high level procedures as well as core framework functions by wrapping the basic libraries low level functionality. In this layer the abstraction techniques have been applied utmost. Here it is possible to adapt the function calls of the layer below to a unified notation of the architecture. Due to this concise abstraction interface it is possible to easily exchange an external library below without adapting the layers above the *core functions*.

The next layer (**module functions**) provides high level functions to be called inside the MMER_Lab modules. The *base class hierarchy* (i.e. module, cable, and framework class) is implemented on this layer in C++ with the advantages of object orientation. Wrappers of framework procedures are also located here using Tcl/Tk. This is the primary layer for the researcher and developer who wants to embed algorithms into MMER_Lab. Due to the object oriented structures in C++ and the high abstraction level, other developers can easily re-use the basic algorithms in their modules.

The top layer (**modules**) consists of module implementations used directly in MMER_Lab. The modules are derived from the base classes of the layer below and implement only function calls, not computations. A Tcl interface to the C++ classes is part of each module. The C/C++ code is stored in shared libraries (.dll/.so) and loaded on demand at runtime.

2.4.2 Design Decisions

Modularity, cross platform usability and performance issues led to several crucial decisions which have been made in the design phase of MMER_Lab. With these three requirements in mind a few existing software systems have been examined regarding the implementation possibilities of video processing algorithms.

ParleVision [95] is a framework for computer vision programming on Windows only. This framework allows to build processing chains, based on filters which support modularity. It has a very good performance in sequential processing but does not support parallelization in any way.

The two other systems, Fermus [74] and Smartflow [77] for network distributed computing applications are process-based utilizing IPC techniques as socket communication and shared memory. These systems, designed to work on networks, support modularity and the usage of multiple processors very well. However, context switching on standalone computers using multi-tasking decreases the performance compared to multi-threaded architectures. This is because multi-tasking systems operate within their own virtual address space. Processes are protected by the operating system from interference by other processes. A user process can not communicate with another process unless it makes use of IPC mechanisms.

Passing the data between computers also comes with network latencies which can affect real time high bandwidth computations such as video processing. The main advantage of these systems is the enormous computing power of a cluster versus the network latency and intercommunication time.

A multi-threaded implementation on a multi-core computer addresses this flaws. Multiple processors are used to distribute the computing tasks and the hardware architecture provides a high-performance platform for exchanging data between the processors. Data can be passed using pointers and mutexes between threads. The context switch of a thread-ed system is lightweight, since the threads all run in a single process and share the same state.

MMER_Lab was chosen to be implemented in this way providing performance and using an API for modularity and a carefree application of multi-threading. If desired, the framework can be extended to work in a distributed network environment by using a middle-ware as Smartflow.

Our multi-threaded approach also supports the tight integration of graphic processors into the framework. With *MMER_GPU* and MMER_Lab GPUs and CPUs can perform different computations in parallel. (gpu.mmer-systems.eu)

2.5 Application Examples and Evaluation

To demonstrate the capabilities of the framework, we instance a concrete video processing and pattern recognition application. It performs the training of a Support Vector Machine (SVM) classifier with feature vectors provided by an Active Appearance Model (AAM) for the automatic recognition of four facial expressions (e.g. neutral, smile, frown, scream) based on a training set of example images. (Fig. 2.2)

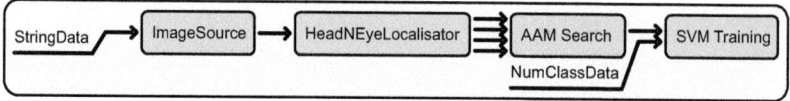

Figure 2.2: Training a SVM with AAM Parameter Data

An 'image source' module is directly connected to a 'head and eye localization' module. The output of the localizer provides a set of four signals (image frame, head region, left-/right-eye position). These dataflows are connected directly to the AAM module for an improved initialization. The output of the AAM module contains the facial feature vector obtained via AAM analysis. This vector data is connected with the first port of the 'SVM' module. The second port is used for the class identifier of the corresponding facial expression.

To begin the SVM training process the system requires two items: the filenames of the training images (for the image source) and their corresponding class assignment (for the SVM module). After the initialization of all modules we use a script to fetch directory content, parse the filenames (which contain the class information) and feed both modules with this data (filename/class pairs).

Here the synchronization abilities become apparent: We provide one filename to the image source (StringData *at the head of the chain*), and one class to the SVM (NumClassData *at the end of the chain*) within one step. The 'NumClassData' cable buffers the information until the SVM module gets a facial feature vector from the AAM. Due to the fact that the SVM-module is built to pull exactly one instance from *each* of the input cables, it is ensured that the class identifier matches with the AAM feature vector of the correct image file.

A simpler application setup was used for evaluation purposes (Tab. 2.1). We use a serial module chain, composed of a video source, an adaptive haar-wavelet head localizer and a display to measure the frame rate (fps) of the localizer algorithm ('one'). This setup is extended to have two localizers in serial ('serial') or in parallel with two displays ('par'). We compare the MMER_Lab system setup using multi-threading (one process, 3-5 threads) with a 'direct' implementation in OpenCV (one process) and an implementation using the NIST Smartflow ('SF') system [77] (3-5 processes). All tests ran on a 320x240 test video. For the series 'one' we used a single core AMD 3700+, 2GB RAM computer with Linux Fedora Core 5 (FC5). The series 'serial' and 'par' were executed on a dual core Intel Pentium4 3.0 GHz, 1GB RAM and FC5. The results show that even for this simple setup the performance of a serial system increases already by one third due to the MMER_Lab based implementation. In comparison with Smartflow, MMER_Lab achieved a higher performance throughout the evaluation, due to the minimized framework overhead of our concept.

System	Single Core			Dual Core		
	one	serial	par	one	serial	par
Direct	59.38	34.84	31.16	51.47	27.00	32.26
MMER	56.40	32.73	30.75	54.2	36.16	31.20
SF	54.63	30.69	29.17	48.72	35.36	30.10
Direct	100%	100%	100%	100%	100%	100%
MMER	94.9%	93.9%	98.7%	105.3%	133.9%	96.7%
SF	92.0%	88.1%	93.6%	94.6%	130.9%	93.3%

Table 2.1: Evaluation results

2.6 Conclusion

When developers strive to exploit the parallel computational power of modern hardware, it is mandatory to separate systems and its components into multiple threads. Since threaded implementation constitutes an own field in software development and requires considerable experience in programming, we built up MMER_Lab as framework to provide a multi-threading environment where the threading and all its difficulties are encapsulated behind interfaces. Due to the module concept of MMER_Lab with the necessity to define interfaces after all, the re-usability of modules is ensured and the system development process within research teams is constructively supported by our framework. MMER_Lab is especially appropriate for systems with high data flow between components, since it allows the handover of memory pointers on whatever kind of data structure. MMER_Lab is in our daily application for development, evaluation and demonstration of systems from the areas of image and video processing or pattern recognition. This validates, that the actualized concepts of MMER_Lab constitute a practical exemplar for generic and flexible environments to identify new design paradigms and to be prepared for the highly parallel hardware architectures of the near future.

Chapter 3

Object Localization with AdaBoost Variants on Haar- and Gabor-Wavelet Features

Many face analysis systems exist which perform accurate face analysis based on manual initialization, demanding for the user to manually set landmark points around and/or in the face [3]. This requirement disqualifies any system from real-life application. When it comes to *automatic* face analysis, as it is the target for a system like *FEASy*, a precise automatic localization of the face is mandatory. We therefore developed and implemented algorithms for a robust localization of faces and eyes and made it available as a module in MMER_Lab (see chapter 2). This module is also a crucial part of the FacE Analysis System (see section 1.2).

The localization of the face and the eyes is carried out in two steps: First the face is localized by means of a bounding box around it. Hence the algorithm localizes both eyes within the face bounding box.

The localization of objects in general or faces and eyes in our case is confronted with the known problems of a great variance in the appearance due to differences in ethnic origin, age, gender, rotation, scale, illumination, and camera properties (see section 1.1).

One approach to this problem is the extraction of an extremely high number of local features from an image and performing a binary classification in "object present at current position" or not. As local features we applied wavelet filters of two kinds, Haar-like [115] and Gabor [66] as trade-off between computational complexity and the capability to characterize the image texture. It can be assumed that the presence of a specific object leads to a characteristic response of specific filters in a certain layout. Hence, the detection of such pattern shall be recognized via the AdaBoost algorithm [41] which samples the image at various positions with windows of variable size in order to cope with scaling effects. In each sample window a set of filter responses is evaluated applying a simple classifier such as Decision Stump separately on each filter. In several cascades the likelihood of the presence of an object is measured. The set of filters, their specific responses, and the cascades are learned during an extensive training phase with thousands of example faces and non-faces. To the well known algorithm of Viola and Jones [115], which is sketched above, we added the Gabor-Wavelet features and examined various other single feature classifiers instead of the Decision Stump. The developments and conducted experiments are described in detail in [124]. It turned out that Gabor-Wavelet features lead to a significantly increased localization accuracy of eyes and that a 1D clustering into 16 adaptive areas is superior to the Decision Stump without implicating an unreasonably increase of training effort. Furthermore, this approach can be applied to the visual distinction of few strong facial expressions.

The subsequent sections gives a short introduction to Haar- and Gabor-Wavelet features, the AdaBoost algorithm and variants, and the performed evaluations.

3.1 Haar-like and Gabor-Wavelet features

3.1.1 Haar-like features

The first wavelet reported in literature was suggested from the German mathematician Alfred Haar. It constitutes the simplest mother wavelet, since it consists in just one step. Its equation can be written as

$$\Psi(x) = \begin{cases} 1 & \text{for } 0 \leqslant x < \frac{1}{2}, \\ -1 & \text{for } \frac{1}{2} \leqslant x < 1, \\ 0 & \text{else.} \end{cases} \tag{3.1}$$

Further is

$$\int_{-\infty}^{\infty} \Psi(x)dx = 0, \int_{-\infty}^{\infty} |\Psi(x)|^2 dx = 1 \tag{3.2}$$

The Fourier Transform of ψ_{Haar} is obtained from

$$\hat{\Psi}(\alpha) = \frac{1}{\sqrt{2\pi}} \left(\int_0^{1/2} e^{-i\alpha x} dx - \int_{1/2}^{1} e^{-i\alpha x} dx \right) \tag{3.3}$$

$$= \frac{i}{\sqrt{2\pi}} \frac{\sin^2(\alpha/4)}{\alpha/4} e^{-i\alpha/2}. \tag{3.4}$$

The plot in figure 3.1 visualizes the band-pass character of Haar-Wavelets since the Fourier Transform is zero for $\alpha = 0$ and $\alpha \to \infty$.

In Image Processing band-pass show strong responses on edges with low noise sensitivity. As those Haar-Wavelet responses can be efficiently computed using the integral image [124], they constitute the first choice for object detection tasks.

3.1.2 Gabor-Wavelets

Another adequate feature for object detection is the Gabor-Wavelet or a set of Gabor-Wavelets respectively. This wavelet has physiological relevance in the human vision process which takes place in the primary cortex of the brain. Three

3.1.2 Gabor-Wavelets

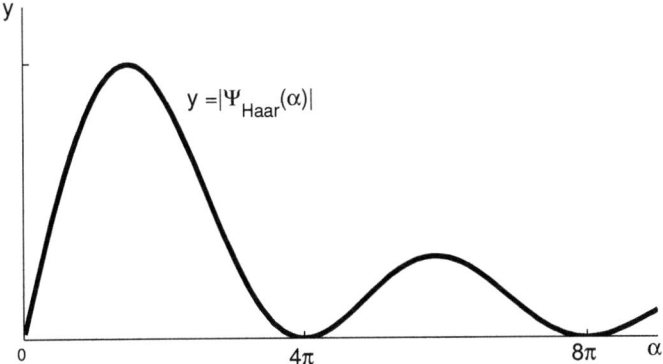

Figure 3.1: 1D Haar-Wavelet spectrum

Figure 3.2: 2D-Haar-like Wavelets filters

kinds of the neurons involved are distinguished: simple, complex, and hypercomplex cells. Those neurons serve as feature extractors while each neuron indicates the presence of a specific geometric feature via micro-impulses. Simple Cells respond to stimuli of a certain orientation and polarity. Complex Cells behave like Simple Cells while they show no separation in inhibiting and inciting areas. Therefore, the position of a stimulus in the receptive field is irrelevant. Hypercomplex Cells respond to stripes, edges, or angles of a certain length which move in a specific direction over a receptive field [92]. Most of the Simple Cells are combined pairwise whereas the cells show different orientations and symmetries. This can be expressed mathematically via the combination of a sine and a cosine function to a complex plane wave form:

$$e^{jkx} = \cos(kx) + j\sin(kx) \quad (3.5)$$

Jones und Palmer [57] showed that the receptive fields of the Simple Cells can be modeled by two-dimensional Gabor-Functions.

Daugmann [28] introduced a two-dimensional Gabor-Filter which provides locality and the spatial frequency with the lowest loss of information at the same time. It can be formulated as follows:

$$G(x,y) = \frac{1}{2\pi\sigma\beta} \cdot \exp\left(-\pi\left[\frac{(x-x_0)^2}{\sigma^2} + \frac{(y-y_0)^2}{\beta^2}\right]\right) \cdot \exp\left(j\left[\xi_0 x + \nu_0 y\right]\right) \quad (3.6)$$

Equation 3.6 describes the multiplication of a Gauss function with a complex plane wave. Hereby, the origin (x_0, y_0) describes the center of the receptive field in the spatial domain and (ξ_0, ν_0) denotes the optimal spatial frequency in the frequency domain. The parameters σ and β determine the standard deviation of the elliptic Gauss function in x- and y-direction [66].

From these considerations a set of Gabor-Wavelets is derived, taking into account the mentioned neuro-physiologic findings:

$$\Psi_i(\boldsymbol{x}) = \frac{\|\boldsymbol{k}_i\|^2}{\sigma^2} \cdot \exp\left(-\frac{\|\boldsymbol{k}_i\|^2 \|\boldsymbol{x}\|^2}{2 \cdot \sigma^2}\right) \cdot \left[\exp(j\boldsymbol{k}_i \boldsymbol{x}) - \exp\left(\frac{\sigma^2}{2}\right)\right] \quad (3.7)$$

Equation 3.7 consists of the following terms:

$\frac{\|\boldsymbol{k}_i\|^2}{\sigma^2}$ Compensation of the frequency dependent power spectrum in natural images

$\exp\left(-\frac{\|\boldsymbol{k}_i\|^2 \|\boldsymbol{x}\|^2}{2 \cdot \sigma^2}\right)$ Gauss function for spatial limitation of the wave

$\exp(j\boldsymbol{k}_i \boldsymbol{x})$ Complex plane wave

$-\exp\left(\frac{\sigma^2}{2}\right)$ Factor for elimination of the constant component reducing the sensitivity to the illumination of the image.

The shape of the filter can be modified via the standard deviation σ of the Gauss function and the vector $\boldsymbol{k}_i$ which is defined by

$$\boldsymbol{k}_i = \begin{pmatrix} k_{ix} \\ k_{iy} \end{pmatrix} = \begin{pmatrix} k_\nu \cos\Theta_\mu \\ k_\nu \sin\Theta_\mu \end{pmatrix} \quad (3.8)$$

We can vary the scale with

$$k_\nu = \frac{k_{max}}{f^\nu} \quad (3.9)$$

and the orientation with

$$\Theta_\mu = \mu \cdot \frac{\pi}{8} \quad (3.10)$$

The parameters used for our purposes follow [69] and [111] and are listed in the following equations.

3.1.2 Gabor-Wavelets

$$\sigma = \pi \tag{3.11}$$

$$k_{max} = \frac{\pi}{2} \tag{3.12}$$

$$f = \sqrt{2} \tag{3.13}$$

$$\nu = 0\ldots 4 \tag{3.14}$$

$$\mu = 0\ldots 7 \tag{3.15}$$

The selection of ν and μ determines one of 5 scales and 8 orientations. Consequently the applied filter set consists in 40 Gabor filters ($i = 1\ldots 40$).

With x_0 and y_0 being the center of a filter kernel equation 3.7 results in

$$\begin{aligned}\Psi_i(x,y) &= \frac{\|\boldsymbol{k}_i\|^2}{\sigma^2} \cdot \exp\left(-\|\boldsymbol{k}_i\|^2 \cdot \frac{(x-x_0)^2 + (y-y_0)^2}{2\cdot\sigma^2}\right) \\ &\quad \cdot \left[\exp\left(j\left(k_{ix}\cdot x + k_{iy}\cdot y\right)\right) - \exp\left(\frac{\sigma^2}{2}\right)\right].\end{aligned} \tag{3.16}$$

Figure 3.3 displays the real part of a Gabor filter kernel whereas figure 3.4 shows the whole family of the 40 Gabor filters applied to feature extraction for object detection.

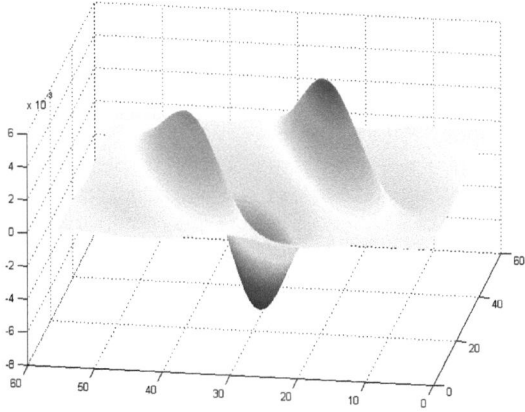

Figure 3.3: Real part of a Gabor filter kernel

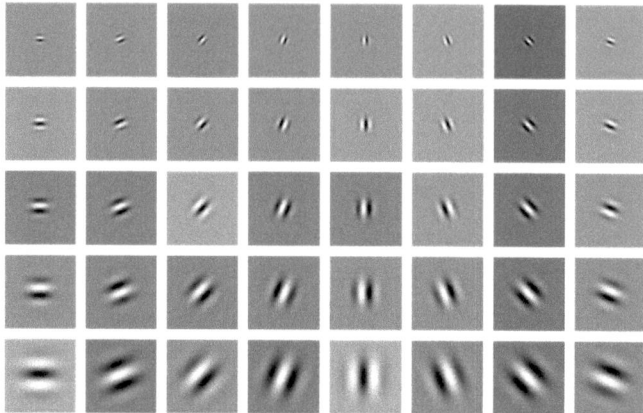

Figure 3.4: Family of Gabor-Wavelet filters

3.2 Feature Selection and Classification with AdaBoost and Variants

The AdaBoost algorithm bases on the idea of *Boosting* which is in general the constructive combination of a set of weak classifiers to a single strong classifier. The advantages of AdaBoost are the low complexity with respect to implementation and runtime plus the very limited number of algorithmic parameters abridging the evaluation procedure for novel problems. Hence, solely the number of training iterations and weak classifiers T need to be adjusted. Furthermore, it is compatible with virtually all classification and learning methods (see 3.2.4).

For object detection we strive to find an ensemble of weak classifiers on basis of single features (Haar-like or Gabor-Wavelets) via AdaBoost. Thereby, the weak classifiers (*abbrev.* 'WC') need to have a classification performance just above random selection. Due to the effective combination of several WCs, a very high detection accuracy can be achieved. Therefore the training of an AdaBoost performs feature selection and learning of a classifier in one. Since we deal with feature spaces of dimensionality $\approx 10^6$ for object detection and apply several thousand example images for training, this algorithm constitutes a very good trade-off between computational complexity and accuracy.

3.2.1 Notation

For the description of the AdaBoost algorithm the following notation is commonly used [41]:

- x: feature vector, $x \in \mathcal{X}$
- y: class assignment $y \in \mathcal{Y} = \{-1; +1\}$

3.2.2 The Standard AdaBoost Algorithm

- M: number of feature vectors in the training dataset
- $(\boldsymbol{x}_1; y_1); \ldots; (\boldsymbol{x}_M; y_M)$: training dataset
- $\omega(i)$: weight of the feature vector $\boldsymbol{x}_i$
- T: number of iterations of the AdaBoost algorithm
- $f(\boldsymbol{x})$: weak classifier (*abbrev.* 'WC')
- $F(\boldsymbol{x})$: strong classifier (*abbrev.* 'SC')

3.2.2 The Standard AdaBoost Algorithm

The basic variant of the AdaBoost algorithm can be described compactly in the following way:

- Start with the weights $\omega_1(i) = 1/M$, $i = 1, 2, \ldots, M$
- Initialize the final strong classifier $F(\boldsymbol{x}) = \mathbf{0}$
- Repeat for all iterations $t = 1, 2, \ldots, T$:
    1. Train the WC $f_t : \mathcal{X} \to \mathcal{Y}$ according to ω_t.
    2. Determine the confidence of the WC $\alpha_t \in \mathcal{R}$.
    3. Update $F(\boldsymbol{x}) = F(\boldsymbol{x}) + \alpha_t \cdot f_t(\boldsymbol{x})$
    4. Update the weights:
    $$\omega_{t+1}(i) = \frac{\omega_t(i)\exp(-\alpha_t y_i f_t(\boldsymbol{x}_i))}{Z_t} \quad (3.17)$$
    while Z_t is a normalization factor leading to a distribution ω_{t+1}.
- Final strong classifier:
$$F(\boldsymbol{x}) = sign\left[\sum_{t=1}^{T} c_t \cdot f_t(\boldsymbol{x})\right] \quad (3.18)$$

The learning algorithm for the weak classifier $f_t : \mathcal{X} \to \mathcal{Y}$ operates in correspondence to the weight distribution ω_t. In the simplest case f_t falls in a binary domain of $\{-1; +1\}$. Hence the weak classifier minimizes the error

$$\epsilon_t = \sum_{i | f_t(\boldsymbol{x}_i) \neq y_i} \omega_t(i) \quad (3.19)$$

Once the rule of the weak classifier f_t is determined, the parameter $\alpha_t \in \mathcal{R}$ indicates the confidence of f_t. For binary problems α_t is adjusted to

$$\alpha_t = \frac{1}{2}ln\left(\frac{1-\epsilon_t}{\epsilon_t}\right). \quad (3.20)$$

Likewise, non-logarithmic confidence measures provide for a higher numerical stability.

The distribution ω_t is updated by decreasing the weight of correctly classified training vectors and increasing the weight of erroneously classified training vectors according to the optimal WC of the current iteration. The following equation describes this methodology.

$$\omega_{t+1}(i) = \frac{\omega_t(i)\exp(-\alpha_t \cdot y_i \cdot f_t(\boldsymbol{x}_i))}{Z_t} \qquad (3.21)$$

Thus, the AdaBoost sets the focus on those vectors hard to classify for subsequent training iterations. After the update, the weights are normalized by the cofficient Z_t:

$$\sum_{i=1}^{M} \omega_{t+1}(i) = 1. \qquad (3.22)$$

The resulting strong classifier consists in the weighted sum of T weak classifiers, whereas α_t constitutes the weight of the WC f_t.

$$F(\boldsymbol{x}) = sign\left(\sum_{t=1}^{T} \alpha_t f_t(\boldsymbol{x})\right) \qquad (3.23)$$

3.2.3 Gentle AdaBoost

Various variants and extensions to the standard AdaBoost have been introduced. One of the first variant constitutes the Discrete AdaBoost [42]. Hereby, a constant represents the confidence of a WC during the update of the training vectors. This further enhances the weight of false classified instances.

Another variant is the Real AdaBoost [122]. Unlike for the Discrete AdaBoost the estimation of the WC confidence is refined herein. The confidence is determined on the training set for both classes separately. The adaptation of the training sample weights is performed dependent on the strength of the confidence. When a sample is correctly classified but with a low confidence, for instance, its weight is less decreased. This leads to a way better adaptation of the weight distribution than with the Discrete AdaBoost.

Finally, the Gentle AdaBoost constitutes an extension of the Real AdaBoost. Here, so called confidence rated predictions are determined instead of single weights for the current weak classifier f_t. Thus sign($f_t(\boldsymbol{x})$) [99] denotes the class and $|f_t(\boldsymbol{x})|$ denotes a confidence measure of the decision for the class. Consequently, one advantage of the Gentle AdaBoost is the reduced influence of outliers due to application of adaptive Newton-Stepping during the assembly of the strong classifier [42].

The update of the strong classifier $F(\boldsymbol{x})$ is explained via

$$f_t(\boldsymbol{x}) = P_\omega(y = 1|\boldsymbol{x}) - P_\omega(y = -1|\boldsymbol{x}) \text{with } f_t \in [-1, +1] \qquad (3.24)$$

Due to the limitation of the domain of f_t numerical instability can be avoided. In several practical investigations the Gentle AdaBoost turned out to be equal or often superior to the other mentioned AdaBoost variants [42]. Therefore, the Gentle AdaBoost was applied in this work for the detection of heads and eyes throughout.

The major difference lies in the process of assembling the final strong classifier:

$$F(\boldsymbol{x}) = \sum_{t=1}^{T} f_t(\boldsymbol{x}) \qquad (3.25)$$

whereas $F(\boldsymbol{x})$ denotes the assigned class and $|F(\boldsymbol{x})|$ the corresponding confidence.

3.2.4 Weak classifiers

Still the question of learning a weak classifier has not been addressed so far. While usually the approach of Decision Stump (*abbrev.* 'DS') classification is applied [42], we investigated several other low-level 1D-classification methods in [124]. Evaluations showed that the DS has difficulties especially for eye localization and was thus replaced by a clustering technique leading to multiple thresholds per feature. The other methods covered interval classifiers with fixed and adapted thresholds determined by clustering techniques. For details please refer to [124].

The Decision Stump approach operates on a single feature, i.e. it deals with a one-dimensional feature space. A DS is described by the parameter set $(k, v, c_{left}, c_{right})$. The classification error is minimized with the threshold v on the space of feature k and provides a confidence for the domain below (c_{left}) and for the domain above (c_{right}) the threshold. The sign of the confidence values indicates the assigned class in a two-class problem.

Let y denote the class label, $\boldsymbol{x}_k$ the value of feature k, j the current positive class, v_j the threshold for class j, and $\omega(i,j)$ the weight of the training vector number i in respect of the class j. Hence, the Decision Stump is explained by

$$n_{11}(j) = \sum_{i=1}^{M} \omega(i,j) \cdot I\{(y_i = j) \wedge (\boldsymbol{x}_k \leqslant v_j)\} \qquad (3.26)$$

$$n_{21}(j) = \sum_{i=1}^{M} \omega(i,j) \cdot I\{(y_i \neq j) \wedge (\boldsymbol{x}_k \leqslant v_j)\} \qquad (3.27)$$

$$(3.28)$$

$$n_{12}(j) = \sum_{i=1}^{M} \omega(i,j) \cdot I\{(y_i = j) \wedge (\boldsymbol{x}_k > v_j)\} \qquad (3.29)$$

$$n_{22}(j) = \sum_{i=1}^{M} \omega(i,j) \cdot I\{(y_i \neq j) \wedge (\boldsymbol{x}_k > v_j)\} \qquad (3.30)$$

with the binary function I:

$$I\{Decision\} = \begin{cases} 1 & \text{if decision correct} \\ 0 & \text{if decision false} \end{cases} \qquad (3.31)$$

In consideration of

$$\sum_{u=1}^{2}\sum_{v=1}^{2} n_{uv}(j) = 1, \forall j. \qquad (3.32)$$

the following equation determines the classification error of the Decision Stump.

$$\epsilon_j = \min\{n_{11}(j) + n_{22}(j), n_{12}(j) + n_{21}(j)\} \qquad (3.33)$$

The accumulated Error of a feature over all classes results in

$$\epsilon = \sum_{j=1}^{J} (\min\{n_{11}(j) + n_{22}(j), n_{12}(j) + n_{21}(j)\}) \qquad (3.34)$$

The probabilities for a correct classification in respect to the threshold are saved as confidence values of the classifier. For each threshold two confidences exist which are determined by

$$c_{left}(j) = \begin{cases} \frac{n_{11}(j)}{n_{11}(j)+n_{21}(j)} & \text{if } n_{11}(j) + n_{22}(j) > n_{12}(j) + n_{21}(j) \\ -\frac{n_{21}(j)}{n_{11}(j)+n_{21}(j)} & \text{else.} \end{cases} \qquad (3.35)$$

$$c_{right}(j) = \begin{cases} -\frac{n_{22}(j)}{n_{12}(j)+n_{22}(j)} & \text{if } n_{11}(j) + n_{22}(j) > n_{12}(j) + n_{21}(j) \\ \frac{n_{12}(j)}{n_{12}(j)+n_{22}(j)} & \text{else.} \end{cases} \qquad (3.36)$$

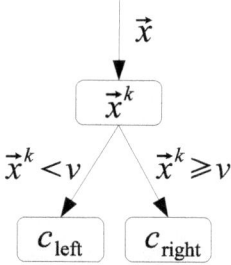

Figure 3.5: Structure of the threshold classifier

Figure 3.6 illustrates the determination of the Decision Stump threshold with an exemplary feature value distribution over the training set.

In figure 3.7 the continuation of the single threshold Decision Stump is displayed leading to adaptive interval placement with a more precise measurement of the classification performance of a single feature. This strategy is referred to as SPLIT throughout the evaluation section.

3.2.5 Cascaded AdaBoost Classification

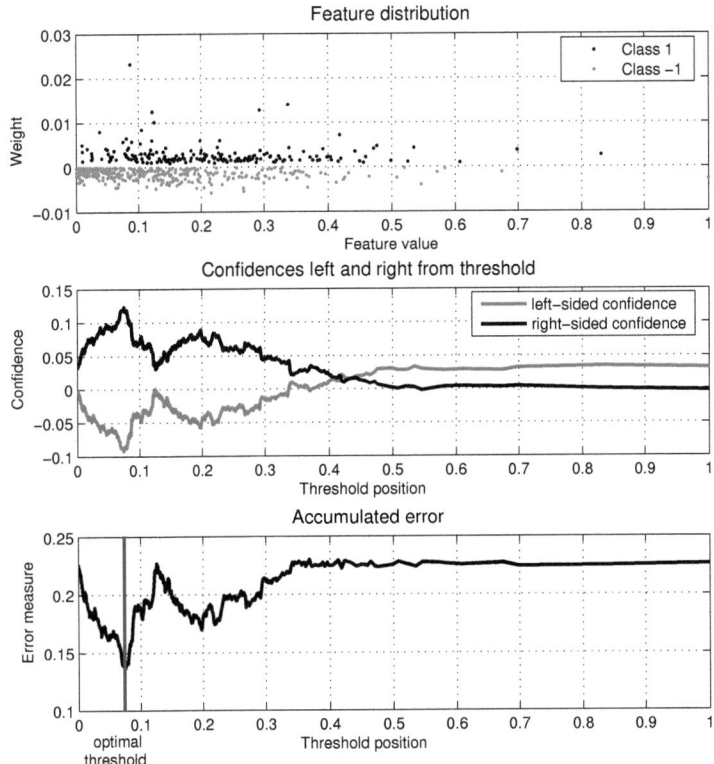

Figure 3.6: Exemplary threshold determination

3.2.5 Cascaded AdaBoost Classification

In order to increase the computational efficiency of object detection tasks, the strong classifier is divided in several stages or cascades. These cascades are trained separately with the target to reject as many as possible sample windows which do not contain the queried object. The positive objects however shall pass all stages and remain as object detections finally (see figure 3.8). Thus, all samples that can be dropped in the first stage do not have to be processed in the subsequent stages which immensely reduces the computational complexity to a fraction. Only for the few positive classified samples, all features and stages have to be computed and evaluated. In the training phase the stages on second or later position are trained only with those samples which have been assigned to the positive class by the directly preceding stage. This covers the actual positive samples plus the false positives which still passed the previous cascades.

A single stage (see figure 3.9) consists in a trained classifier designed to reject

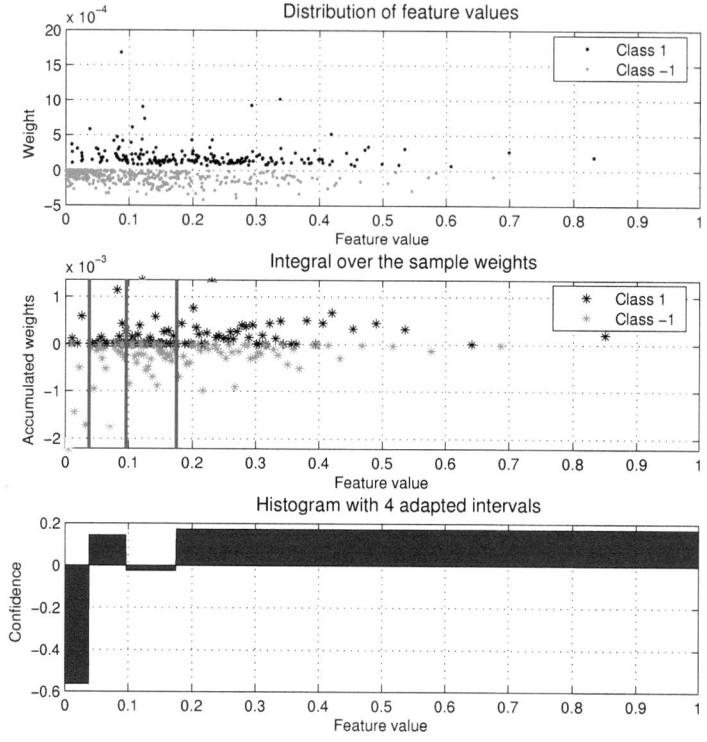

Figure 3.7: Determination of intervals with the SPLIT strategy

e.g. at least 50% of the false positive samples. At a cascade of ten stages the false positive rate falls at least below $\frac{1}{1024}$ while the detection rate holds at $0.995^{10} = 0.95$.

Although each single stage classifier can base on any classification method, we applied the AdaBoost algorithm to create a high-performance face and eye localizer with respect to accuracy *and* speed.

3.3 Evaluation of Localization Performance

This section presents the results of our face and eye localization algorithm which bases on the technologies mentioned above in 3.2. Apart from the databases and evaluation setup, related findings from our research work in this area are presented in short.

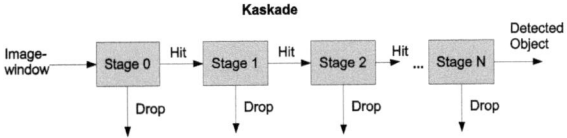

Figure 3.8: Structure of cascades for object detection

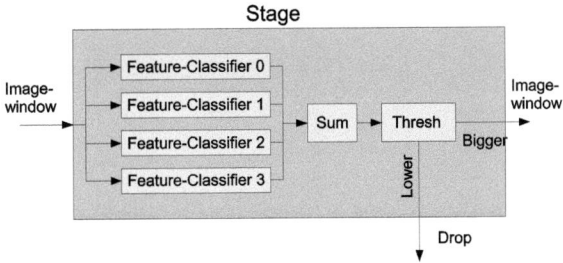

Figure 3.9: Single stage of a cascade

3.3.1 Databases

We included several image databases in order to obtain a rich set of faces, eyes, and non-faces. Approximately 10.000 sample windows were extracted from the do-it-yourself store picture set of the NI-Face (see section 6.2.2) database as negative face examples. This set is especially valuable since it shows many variants of heavily cluttered background. A set of 5052 faces and accordingly 10104 eyes were extracted from the face images of the NI-Face, the AR (see section 6.2.1), and the FERET database [93].

The latter contains a large set of faces from many different persons in a balanced arrangement of gender, ethnic origin, facial expression, and head poses. The database consists of overall 14051 gray scale images and was collected for the training and evaluation of face recognition algorithms.

3.3.2 Head and Eye Localization Results

For initialization of the face analysis with statistical shape and texture models (see section 1.2) primarily a precise eye localization is required. As target localization accuracy, a deviation of less than 5 pixels of Euclidean distance to the eye center shall be achieved. This constitutes the basis for the performance evaluations in table 3.2. The appearance of faces show a significantly higher complexity compared to eyes. Therefore, the localization of faces can be performed much more efficiently and accurately.

The described approach of a cascaded AdaBoost classifier on Haar-like features performs a head recognition with 100% detection rate of faces and single

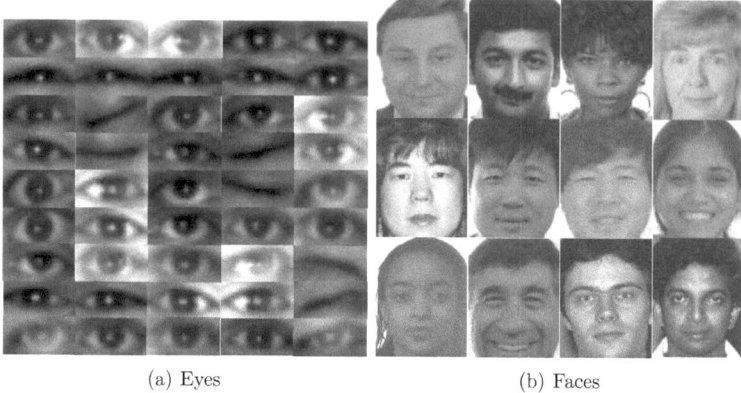

(a) Eyes (b) Faces

Figure 3.10: Images of the FERET database and positive material for the eye localization

false detections in 2 out of 118 pictures of the non-face corpus of the NI-Face database. Due to this performance, in our system the initialization bases on an eye localization which requires a preceding face localization. The latter provides a bounding box of the detected face with width w_{bb} and height h_{bb}.

During the eye localization process the detected face bounding box is scaled to $w_{bb} \times h_{bb} = 90 \times 120$ pixels in order to bring the eyes to the specified size of 20×10 pixels. This bounding box is divided in 4 horizontal stripes of height $1/4 h_{bb} = 30$. The eyes are assumed to be in the second stripe from the top.

Within this stripe the eyes are localized with a specially trained AdaBoost in six cascades based on Haar-like *and* Gabor-Wavelet features (see section 3.1). It turned out that a separated training and classification of left and right eyes leads to a lower accuracy than a combination of both types. This might be due to high similarity of both eyes in the features space and the loss of positive training material when separated. All training samples are resized to 20×10 pixels. The rest of the face bounding boxes in the training material serves as negative data samples. Each stage of the AdaBoost cascade is required to achieve a false acceptance rate lower than 50% and a positive detection rate of minumum 99.5%.

The evaluation of the created cascade is tested on a separate evaluation dataset of the FERET-Database with 908 frontal faces. The position of a localization is determined when at least 8 windows with an absolute translation of less than 6 pixels agglomerate at a certain point in the search area. The eye position constitutes the centroid of the average of the 8 or more windows.

3.3.3 Feature selection

The number of selected features for eye localization is relatively high when training an AdaBoost cascade. In order to measure the (computational) efficiency

3.3.4 Localization Performance

	H		G		H+G		
	overall #	average #	overall #	average #	overall #	average #	G#
STUMP	796	53,6	1418	41,1	1216	37,8	131
LUT 8	1041	24,6	4163	104,8	753	21,5	41
LUT 16	705	19,7	1601	45,9	533	19,2	85
LUT 32	520	16,01	880	36,5	427	16,5	53
SPLIT 8	1450	32,7	1586	84,2	989	34,3	49
SPLIT 16	757	26	921	47	683	25,9	54
SPLIT 32	603	19,2	742	30,8	456	18,3	104

Table 3.1: Number of required features (overall #) and average number of computed features per sampled window on the test dataset (average #), as well as the number of Gabor features at application of both feature types (H+G) (G #)

however it is more expressive to compare the average number of computed features during the localization on a dataset. Table 3.1 shows that the overall number of selected features is not conform with the number of computed features. A highly complex classifier, such as LUT8, a look-up table with 8 bins on basis of Haar-like features, requires 1041 selected features. The average number of evaluated features however is just 24.6.

3.3.4 Localization Performance

	H		G		H+G	
	HR (%)	average #	HR(%)	average #	HR(%)	average #
STUMP	95,15	53,6	96,92	41,1	97,14	37,8
LUT 8	98,02	24,6	94,71	104,8	98,35	21,5
LUT 16	98,24	19,7	95,36	45,9	98,13	19,2
LUT 32	97,79	16,0	94,93	36,5	97,47	16,5
SPLIT 8	98,02	32,7	97,03	84,2	98,13	34,3
SPLIT 16	98,46	26	96,70	47,0	98,46	25,9
SPLIT 32	97,36	19,2	96,15	30,8	98,35	18,3

Table 3.2: Hit rate (HR) and average number of computed features per window on the test dataset (average #)

The evaluation of the trained eye localization cascades does not show extreme differences in the hit rate. All methods provide remarkably good results. The essential difference lies in the number of computed features, i.e. the efficiency of the approach. We must not forget to mention that the computation of Gabor wavelet features is up to 20 times computationally more expensive than Haar-like features derived from the integral image. The combination of Haar and Gabor features provides for a slightly increased computation time, but leads to the best recognition results.

	HWK	GWK	H+G
STUMP	219	601	182
LUT 8	185	2219	153
LUT 16	164	1743	193
LUT 32	193	1452	215
SPLIT 8	177	3797	225
SPLIT 16	165	2004	189
SPLIT 32	154	3070	238

Table 3.3: Average duration of the eye localization per face (in ms) regarding the used weak classifier

For the problem of eye localization it becomes obvious that the Decision Stump is too simple for modeling the present feature spaces. The best results could be achieved with the Split method. With the developed approach, 98.5% of the eyes are located with less than 5 pixels deviation from the center with in average just 25.9 computed features. This documents a very fast and accurate eye localization technique.

Apart from object detection we successfully applied a multi-class extension of the Gentle AdaBoost to facial expression analysis and head pose recognition. Thereby, the four expressions frown, smile, scream, and neutral or the poses frontal, 45, and 90 degrees left and right were distinguished respectively [124, 76]. Although this algorithm is not capable to discriminate objects with just small differences (e.g. frown vs. neutral expression), it constitutes a valuable approach for a granular and computationally cheap recognition of object variants.

3.4 Conclusion

In our FacE Analysis System FEASy (section 1.2) the localization of a face and its eyes serves as necessary initialization of the subsequent analysis based on shape and texture models. Hence, we apply an algorithm following the approach presented by Viola and Jones [115] for the face localization. This localization algorithm is based on a sampling of an image with windows of variable size. From each sample window visual Haar-like wavelet features are extracted. Thereby a Decision Stump as weak classifier operates on single features. These weak classifiers are combined by a Gentle AdaBoost [42] which tries to reject windows without a face at early stages of a cascade. The localization of the eyes runs on a narrowed area within the face region provided by the face localization.

We developed several improvements to the standard Viola-Jones algorithm for a more accurate eye localization. It turned out that the addition of Gabor wavelet features and the replacement of the Decision Stump as weak classifier by an adaptive interval classifier leads to a localization at higher efficiency and smaller spatial deviation. Finally, according to our evaluations 98.5% of the eyes in pictures with frontal human faces can be localized with less than 5 pixel Euclidean deviation from the actual eye center when the face is scaled to a size of 90 × 120 pixels based on the automatic face localization. The software implementation runs more than 5 times real-time on images of double VGA resolution. Therefore, the developed improvements provide a reliable basis of our fully automatic face analysis system.

Chapter 4

The Theory of Active Appearance Models

This chapter gives an introduction to the mathematical and implementation background of Active Appearance Model (*abbrev.* 'AAM') algorithm as it was originally proposed by Cootes et al. [27], however with a unified mathematical notation. Minimal modifications of the algorithm are already introduced here with a special notification, while chapter 5 describes the manifold variations, extensions, and optimization developed during the research for this work.

The approach of AAMs is based on the generation of statistical models of the appearance, i.e. the shape and texture, of a given object class. AAMs base on the assumption that the two-dimensional appearance of an object in a digital image is influenced by several widely independent sources. The target for the generation of an Appearance Model (*abbrev.* 'AM') is to eliminate the variance (see fig. 4.1) induced by the camera view (translation, rotation, scale), illumination conditions, and the capturing device (brightness, intensity). Thus, the AM focuses on the independent modeling of the shape and texture in a first step and finally combines the two sources again to model the influences of shape variations on the texture. Furthermore, an AM allows the synthesis of a variety of instances of the same object class by adjusting a set of scalar AM coefficients. Therefore, AAMs can be applied to the visual analysis of unknow objects of the given class. The analysis is performed via the best possible re-synthesis of the object by solving the optimization problem of the automatic adjustment of the Appearance Model coefficients. The originally proposed AAM algorithm limits the computational effort of the optimization by following a *predicted* gradient. This prediction is obtained from an averaged sampling of the search space which is performed a-priori under the assumption that the search space is similar for all targeted objects. The optimized coefficient values constitute a low-dimensional representation of the analyzed object for further processing, e.g. in Pattern Recognition tasks. In context of this work the targeted visual objects are human faces which shall be analyzed with respect to the gender, age, identity, head pose or facial expression of a person (see chapter 6).

Figure 4.1: Variance in the appearance of human faces [1]

4.1 Preparation of Training Data

4.1.1 Alignment and Normalization of Landmarks

In the first step of the AM generation, the pure shape information in the training data, i.e. face images, must be extracted.

Let $\mathcal{P}$ be a set of images and $\mathcal{S}$ a set of corresponding shapes which are used to build a *shape model*, a *texture model* and finally a *combined model*.

The training images $\boldsymbol{p}_i \in \mathcal{P}$ with $0 \leqslant i < p$ have to be annotated, producing a set of p corresponding landmark vectors $\boldsymbol{s}_i \in \mathcal{S}$ where $\boldsymbol{s}_i$ is the ith landmark vector defined as the concatenation of all 2D landmark coordinates

$$\boldsymbol{s}_i = (x_0, y_0, x_1, y_1, \ldots, x_{(L/2)-1}, y_{(L/2)-1})^T \quad (4.1)$$

See figure 4.2 for example annotations with $L/2 = 72$ landmarks.

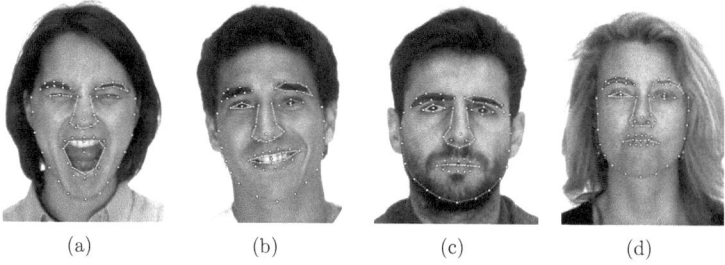

Figure 4.2: Annotation scheme for faces with 72 landmarks

All the meshes defined by the landmarks in the p training images usually face a high deviation in translation, rotation, and scaling. Since a shape is considered as the relative spatial position of all landmark points, the vectors $\boldsymbol{s}_i \in \mathcal{S}$ are aligned to each other and normalized in order to remove Euclidean transformations and scaling and to minimize the variance to the deformation of the shapes.

4.1.1 Alignment and Normalization of Landmarks

The algorithm of the alignment strives to determine a mean shape $\overline{s}$ which minimizes the accumulated Euclidean distance of all landmarks of all shapes to their reference landmark of the mean shape. The initial estimation of the mean shape $\hat{s}$ is usually chosen as one of the training shape vectors s_i and hence updated in an iterative algorithm.

It adjusts the parameters of the similarity transformation T defined as

$$\text{T}: \quad t(x) = \begin{pmatrix} a & -b \\ b & a \end{pmatrix} x + \begin{pmatrix} t_x \\ t_y \end{pmatrix} \quad (4.2)$$

in order to transform the landmarks of s_i so that the square distance between the aligned shape s'_i and $\hat{s}$ is minimized.

Let s be an arbitrary shape vector and $\hat{s}_{[n]}$ the current estimation for the mean shape in iteration n. Define $x_l = (s_{i_{2l}}, s_{i_{2l+1}})^T = (x_l, y_l)^T$ as the lth landmark from the ith shape vector s_i and $x'_l = (\hat{s}_{[n]2l}, \hat{s}_{[n]2l+1})^T = (x'_l, y'_l)^T$ as the lth landmark from the current estimation of the mean shape $\hat{s}_{[n]}$. Then the transformation parameters a, b, t_x and t_y shall minimize the square distance

$$\text{E}(a, b, t_x, t_y) = \sum_{k=0}^{L-1} \|t(x_l) - x_l'\|^2 \quad (4.3)$$

The task can be simplified by translation of all shape vectors into their center of gravity ($t_x = t_y = 0$). In this special case

$$a = \frac{s \cdot \hat{s}_{[n]}}{\|s\|^2} \quad (4.4)$$

$$b = \frac{L}{\|s\|^2} \sum_{l=0}^{L-1} x_l y'_l - y_l x'_l \quad (4.5)$$

is a solution for an arbitrary shape vector s and the current estimation of the mean shape $\hat{s}_{[n]}$. Consequently, the updated estimations $\hat{s}_{n+1}$ converge against the actual $\overline{s}$. The algorithm terminates when $\Delta \hat{s} = |\hat{s}_{[n]} - \hat{s}_{[n-1]}| = 0$.

Finally, we set $\hat{s}_{[n]} = \overline{s}$ and obtain a set $\mathcal{S}'$ of aligned shape vectors s'_i.

Projection into Tangent Space The generation of an AM provides for subspace methods like Principal Component Analysis (*abbrev.* 'PCA') or Non-negative Matrix Factorization (*abbrev.* 'NMF') in order to achieve a high information reduction and a compact description of the training data. However, PCA and NMF try to explain only linear correlations in the data set. The tangent space coordinate transformation eliminates the non-linear properties of the shape set and thus improves the modeling of the shape variance [27].

The projection into tangent space (*abbrev.* 'TS') is simply achieved by applying the following transformation to each aligned shape vector $s'_i \in \mathcal{S}'$:

$$s_i^{TS} = \frac{1}{s'_i \cdot \overline{s}} \quad (4.6)$$

These aligned shape vectors are arranged in the *shape matrix* S with

$$S = \begin{bmatrix} s_0{}^{TS} \mid s_1{}^{TS} \mid \cdots \mid s_{p-1}{}^{TS} \end{bmatrix} \in \mathbb{R}^{L \times p} \qquad (4.7)$$

Still, the *mean shape* $\bar{s}$ is defined as the mean of all shape vectors in S. For improved readability, the term *shape* and the symbol s_i henceforth references the column vectors of S.

4.1.2 Warping

In context of this work, the warping transformation $W(p, s)$ cuts out the inner regions of the shape s from the image p and deforms it to the mean shape $\bar{s}$. This operation is performed in two different occasions within the AAM algorithm:

1. During AM generation: The texture variance of all training images is reduced by the warping of all textures inside the shapes s_i to the mean shape (see section 4.1.3.

2. During AAM coefficient optimization: A shape is placed on an image to be analyzed. For estimation of the quality of the current re-synthesis, the texture inside the shape must also be warped to the mean shape (see section 4.3.1.

The approach of the deformation of a polygon, such as the mesh of a shape, requires a simplification of the problem by dividing the region into a set of triangles $\psi_j \in \ominus$. An inexpensive divide-and-conquer method to calculate a triangulation $\ominus$ for a set of scattered multi-dimensional points is described by Paolo Cignoni et al. in [16]. See figure 4.3 for an illustration.

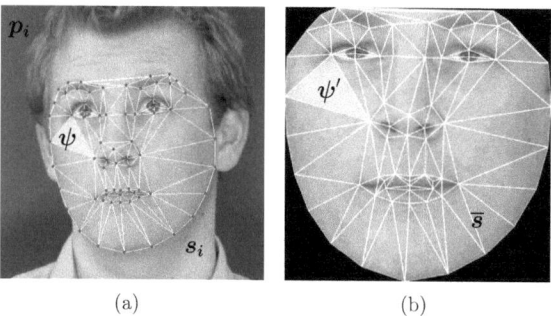

(a)   (b)

Figure 4.3: (a) Source image with source shape s_i and triangle ψ - (b) position and shape of the target triangle ψ' in the mean shape $\bar{s}$

Eventually, the image warping consists in the mapping of textures within corresponding triangles of a complex polygon.

4.1.3 Normalization of Textures

Let M be a transformation operation that maps an input triangle $\psi_j \in \ominus$ to the corresponding transformed triangle ψ'_j in the destination area. When M is applied to a triangle, in a first step it is applied to each of the vertices of the triangle defined by the respective landmark coordinates. The pixels in the image $p_i \in \mathcal{P}$ covered by the source triangle are then projected into the destination region accordingly, using linear interpolation.

The transformation M is now applied to all triangles $\psi_j \in \ominus$. The final result of this operation is a deformed image in the destination area. If all landmarks are placed correctly on the faces and the warping transformation $W(\boldsymbol{p}, \overline{\boldsymbol{s}})$ is applied to all p images, the facial features (eyes, brows, nose, mouth, etc.) are warped always to the same location within the mean-shaped texture. This is crucial for the generation of a statistical model of the texture variation as described in 4.2.2. Foley et al. give a comprehensive explanation of the theory of texture mapping and pixel transformation in [40]. On the one hand the task of triangle-based warping shows an extreme computational complexity due to the extremely high number of pixels and color values and their linear interpolation during texture mapping. For instance, the warping into a texture of a mean face with maximal x- and y-dimensions of 128 × 128 requires the computation of approx. $40 \cdot 10^3$ values assuming three color channels per pixel. On the other hand, warping is one of the highly optimized core operations of modern graphics hardware. Thus, our implementation of Active Appearance Model exploits this computational power by performing the warping task on the GPU. More details will be given in section 5.4.

4.1.3 Normalization of Textures

The normalization of the textures from the p training images $\boldsymbol{p}_i$ comprises the warping of all pixel areas within the corresponding shapes $\boldsymbol{s}_i$ to the mean shape $\overline{\boldsymbol{s}}$ plus the normalization to an average brightness and intensity of all training textures.

The texture within the annotated shape $\boldsymbol{s}_i$ of each training image is warped by the warping transformation W to fit the mean shape $\overline{\boldsymbol{s}}$. This produces a shape-free representation $\boldsymbol{t}_i$, $0 \leq i < p$ of the original texture in the training image.

$$\boldsymbol{t}_i = W(\boldsymbol{p}_i, \boldsymbol{s}_i) \tag{4.8}$$

The warping transformation W is discussed in section 4.1.2.

For the generation of the texture model, we store the obtained set of textures $\boldsymbol{t}_i \in \mathcal{T}$ as vectors column-wise in the *texture matrix* $\boldsymbol{T}$

$$\boldsymbol{T} = [\boldsymbol{t_0} \mid \boldsymbol{t_1} \mid \cdots \mid \boldsymbol{t_{p-1}}] \in \mathbb{R}^{c\chi \times p} \tag{4.9}$$

where c is the total number of pixels in each texture and χ is the number of (color-) channels. For grayscale textures the number of channels is $\chi = 1$, for interleaved RGB color textures it is $\chi = 3$.

In order to eliminate the texture variance caused by brightness and contrast disparities (see fig. 4.1) and for the computation of a mean texture $\bar{t}$, an iterative image alignment algorithm is applied.

The *brightness* b_i of a texture t_i describes the mean over all color values of all pixels:

$$b_i = \frac{1}{c\chi} |t_i| \qquad (4.10)$$

For further processing each texture t_i is freed from its specific brightness level b_i resulting in texture vectors $t'_i \in \mathcal{T}'$.

Beside the additive brightness, digital images differ in contrast disparities captured by the multiplicative scaling factor describing the *intensity* m. Therefore, a texture t of a training image can be represented by its mean-free color values t', the brightness, and the intensity as follows:

$$t'_i = (t_i + b_i) \cdot m \qquad (4.11)$$

In order to determine a mean texture $\hat{t}$ over all training textures, the brightness-free texture vectors are iteratively scaled and aligned. An initial mean texture $\hat{t}_{[0]}$ can either be a specific texture from within $\mathcal{T}$ or calculated as the mean over all textures $t_i \in \mathcal{T}$, $0 \leqslant i < p$ by evaluating

$$\hat{t}_{[0]} = \frac{1}{p} \sum_{i=0}^{p-1} t_i \qquad (4.12)$$

Then the estimated mean texture $\hat{t}_{[0]}$ is freed from its brightness and scaled to unit length. The actual normalization of the textures is an iterative adjustment of aligned textures $t'_i \in \mathcal{T}'$ on the basis of the current estimation for the mean texture $\hat{t}_{[n]}$ and then calculating $\hat{t}_{[n+1]}$ for the next iteration.

At the beginning of each iteration, every texture $t'_i \in \mathcal{T}'$ is made mean-free and scaled to the length of $\hat{t}_{[l]}$ using a value $a_{[l]}$:

$$a_{[l]} = \frac{1}{\hat{t}_{[l]}.t'_i} \qquad (4.13)$$

The re-estimation of the new mean texture $\hat{t}_{[l+1]}$ is calculated on the modified texture vectors by evaluating

$$\hat{t}_{[l+1]} = \frac{1}{p} \sum_{i=0}^{p-1} t'_i \qquad (4.14)$$

The iterative algorithm will continue as long as the mean square error between the old estimation for the mean texture $\hat{t}_{[l]}$ and the new estimation $\hat{t}_{[l+1]}$ is greater than a given error threshold. In this case we can assume that the algorithm converged and that $\hat{t}_{[l+1]} \approx \bar{t}$ for an l large enough. Finally we get a set $\mathcal{T}'$ of normalized and aligned texture vectors t'_i. For improved readability, the term *texture* and the symbol t_i henceforth references the aligned texture vectors stored column-wise in the matrix T.

Thus, the influence of the overall lighting conditions and camera sensitivity is widely reduced. This allows for a statistical modeling which focuses on the actual variance in the texture of an object class widely independent from the capturing scenario.

4.2 Generation of an Appearance Model

4.2.1 Shape Model

The *shape model* is built by applying a PCA to the shape matrix S, i.e. an Eigenvalue Decomposition of the Covariance Matrix over all shapes s_i. The obtained Eigenvectors constitute the *shape basis* Φ_s, whereas basis vectors are sorted in descending order of their corresponding Eigenvalue λ_{si}. Information reduction is achieved by only selecting the top μ_s "most important" basis vectors, discarding those which correspond to principal axes bearing few variance of the data. Evaluations showed throughout that the remaining basis vectors should explain 98% of the total variance. Since the size of the Eigenvalue λ_{si} indicates the variance explained by the ith Eigenvector, μ_s can easily be determined by

$$\frac{\sum_{i=0}^{\mu_s-1} \lambda_{si}}{\sum_{i=0}^{L-1} \lambda_{si}} \overset{!}{\geqslant} 0.98 \qquad (4.15)$$

The same method is applied for the texture and combined model. A new shape

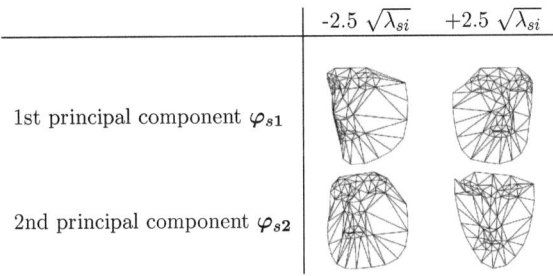

Figure 4.4: Effect of the first shape model components

s can be synthesized by the linear combination

$$s = \overline{s} + \Phi_s h_s \qquad (4.16)$$

whereas h_s contains the *shape coefficients* that control the deformation of the shape model. Note that a zero coefficient vector relates to the mean shape $\overline{s}$. As Φ_s defines an orthonormal basis, the new representation h_{si} of the known shape s_i in the PCA space can be obtained by

$$h_{si} = \Phi_s^T (s_i - \overline{s}) \qquad (4.17)$$

4.2.2 Texture Model

The *texture model* is built by applying another PCA to the texture matrix $\boldsymbol{T}$. Note that the covariance matrix $\frac{1}{c\chi}\boldsymbol{T}\boldsymbol{T}^T$ would consist of $c\chi$ rows and columns which is highly problematic in respect of memory consumption and runtime complexity of the Eigenvalue Decomposition. However, $\boldsymbol{T}$ fulfills the same conditions as known from the Eigen-Face approach [113]. The Eigenvectors of the covariance matrix $\frac{1}{c\chi}\boldsymbol{T}\boldsymbol{T}^T \in \mathbb{R}^{c\chi \times c\chi}$ can also be obtained by determination of the Eigenvectors of $\frac{1}{p}\boldsymbol{T}^T\boldsymbol{T} \in \mathbb{R}^{p \times p}$. Those Eigenvectors $\boldsymbol{\Phi}'_t$ require an additional projection back in the correct space applying an orthonormal normalization to $\boldsymbol{\Phi}_t = \boldsymbol{\Phi}'_t \boldsymbol{T}$. A detailed description of this well-known memory and runtime optimization for PCA analysis is given in [113]. The result is a *texture basis* $\boldsymbol{\Phi}_t$, whereas basis vectors are sorted in descending order of the corresponding Eigenvalues $\lambda_{t0} < \lambda_{t1} < \ldots < \lambda_{t\mu_t-1}$. Again, the first μ_t Eigenvectors $\boldsymbol{\varphi}_t$ are preserved, while the "least important" Eigenvectors of $\boldsymbol{\Phi}_t$ are discarded according to eq. 4.15.

A new texture $\boldsymbol{t}$ can be synthesized in the shape-free space by

$$\boldsymbol{t} = \bar{\boldsymbol{t}} + \boldsymbol{\Phi}_t \boldsymbol{h}_t \tag{4.18}$$

with $\boldsymbol{h}_t$ containing the *texture coefficients* to manipulate the synthesized texture. Note that a zero coefficient vector relates to the mean texture $\bar{\boldsymbol{t}}$. As $\boldsymbol{\Phi}_t$ defines an orthonormal basis, the new representation $\boldsymbol{h}_{ti}$ of the known texture $\boldsymbol{t}_i$ in the PCA space can be obtained from

$$\boldsymbol{h}_{ti} = \boldsymbol{\Phi}_t^T (\boldsymbol{t}_i - \bar{\boldsymbol{t}}) \tag{4.19}$$

4.2.3 Combined Model

In order to finally generate the Appearance Model, shape and texture correlations are recovered from the so far independent shape and texture models. Let $\boldsymbol{c}_i$ be the ith vector which contains the concatenated shape and texture coefficient vectors $\boldsymbol{h}_{si}$ and $\boldsymbol{h}_{ti}$ for each of the $0 \leqslant i < p$ training samples

$$\boldsymbol{c}_i = \begin{pmatrix} \boldsymbol{K}\boldsymbol{h}_{si} \\ \boldsymbol{h}_{ti} \end{pmatrix} \tag{4.20}$$

$\boldsymbol{K}$ is a diagonal matrix of reasonable weights to equalize unit differences between the shape and the texture model. As Cootes and Taylor [27] suggest, a reasonable approach is to set $\boldsymbol{K} = k\boldsymbol{I}$ where k is the ratio of the total intensity variance of the textures to the total shape variance and can be written as

$$k = \frac{\sum_{i=0}^{\mu_t - 1} \lambda_{ti}}{\sum_{i=0}^{\mu_s - 1} \lambda_{si}} \tag{4.21}$$

The vectors $\boldsymbol{c}_i$ form the matrix of concatenated coefficient vectors $\boldsymbol{C} = [\boldsymbol{c_0} \mid \ldots \mid \boldsymbol{c_{p-1}}]$ and is defined as

$$\boldsymbol{C} = \begin{bmatrix} \boldsymbol{K}\boldsymbol{\Phi}_s^T [\boldsymbol{S} - \bar{\boldsymbol{s}}\boldsymbol{1}^T] \\ \boldsymbol{\Phi}_t^T [\boldsymbol{T} - \bar{\boldsymbol{t}}\boldsymbol{1}^T] \end{bmatrix} \tag{4.22}$$

where **1** is the vector containing all ones and $\mathbf{1} \in \mathbb{R}^L$ or $\mathbf{1} \in \mathbb{R}^{cx}$ respectively. Since the shape coefficients h_{si} and texture coefficients h_{ti} are already mean-free so is C.

Another PCA is applied to the matrix C producing the *combined basis* Φ_c, whereas basis vectors are sorted in descending order of the corresponding Eigenvalue λ_{ci}, again discarding the "least important" basis vectors according to eq. 4.15.

A coefficient vector c for the shape and texture models can be synthesized by evaluating

$$c = \Phi_c h_c \tag{4.23}$$

where h_c contains the *AM coefficients*. As the matrix Φ_c can be split into the shape and texture relevant parts Φ_{cs} and Φ_{ct}

$$\Phi_c = \begin{bmatrix} \Phi_{cs} \\ \Phi_{ct} \end{bmatrix} \tag{4.24}$$

it is possible to express a new shape s and texture t directly as function of h_c by combining eq. 4.22 with eq. 4.16 and 4.18:

$$s(h_c) = \bar{s} + Q_s h_c \quad , \quad Q_s = \Phi_s K^{-1} \Phi_{cs} \tag{4.25}$$
$$t(h_c) = \bar{t} + Q_t h_c \quad , \quad Q_t = \Phi_t \Phi_{ct} \tag{4.26}$$

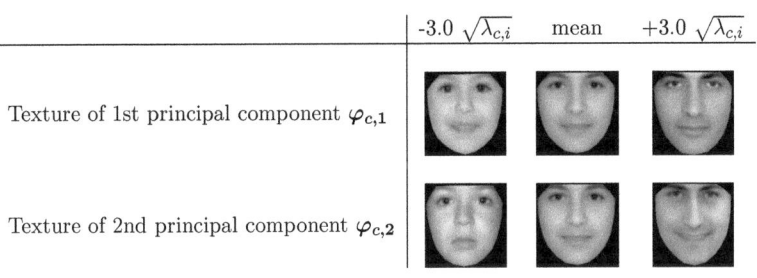

Figure 4.5: Effect on the texture of the first two combined model components of an AM built from the FG-NET Aging Database [1]

4.3 Coefficient Optimization

The application of Active Appearance Models on the analysis of faces in the first place consists in re-synthesizing an unknown face in a video or picture. In other words, the analysis task requires an automatic adjustment of the AM coefficient in order to find a synthesized face which is as similar as possible to the analyzed face. Thus, the AAM coefficient optimization problem can be interpreted as a multivariate, vector-valued optimization task. By minimizing the difference between the re-synthesized face and the original one, a locally optimal coefficient vector can be found by an iterative optimization algorithm. Since transformations in the image plane as well as intensity and brightness have been eliminated during AM training, they have to be integrated again during the optimization process.

4.3.1 Objective Function

This section deals with the objective function which is subject to the multi-variate optimization of the AM coefficients.

Let the error energy E($\boldsymbol{h}$) be defined as

$$\mathrm{E}(\boldsymbol{h}) = \frac{1}{2}\|\boldsymbol{r}(\boldsymbol{h})\|^2 \tag{4.27}$$

where $\boldsymbol{r}(\boldsymbol{h})$ is the difference image between the re-synthesized face and the original face, as described in eq. 4.30. The goal is to minimize E($\boldsymbol{h}$) in respect to an extended coefficient vector $\boldsymbol{h}$ such that

$$\operatorname*{argmin}_{\boldsymbol{h}} \frac{1}{2}\|\boldsymbol{r}(\boldsymbol{h})\|^2 \tag{4.28}$$

The coefficient vector $\boldsymbol{h}$ is obtained by extending the AM coefficient vector $\boldsymbol{h}_c$ by additional *pose* and *texture coefficients*. These are used to optimize the position $(t_x, t_y)^T$ scaling s and rotation r of the shape model in the image plane as well as the intensity m and brightness b of the texture model. Let the extended coefficient vector $\boldsymbol{h}$ be

$$\boldsymbol{h}^T = (t_x \mid t_y \mid s \mid r \mid m \mid b \mid \boldsymbol{h}_c^T) \in \mathbb{R}^v \tag{4.29}$$

Difference image The difference image $\boldsymbol{r}(\boldsymbol{h})$ can be written as

$$\boldsymbol{r}(\boldsymbol{h}) = A \circ W(\boldsymbol{p}, \boldsymbol{s}(\boldsymbol{h}_c)) - \boldsymbol{t}(\boldsymbol{h}_c) \tag{4.30}$$

This equation is essential for all coefficient optimization algorithms. We focus on the minimization of the difference between the two textures $A \circ W(\boldsymbol{p}, \boldsymbol{s}(\boldsymbol{h}_c))$ and $\boldsymbol{t}(\boldsymbol{h}_c)$. Hereby, $A \circ W(\boldsymbol{p}, \boldsymbol{s}(\boldsymbol{h}_c))$ corresponds to the linearized representation of the warped and aligned texture that is gained from the original image $\boldsymbol{p}$. The synthesized shape $\boldsymbol{s}(\boldsymbol{h}_c)$ (see eq. 4.25) and the pose coefficients t_x, t_y, s, and r define a region in $\boldsymbol{p}$. This region is warped from $\boldsymbol{p}$ and into the mean shape $\overline{\boldsymbol{s}}$ using the warping transformation W defined in section 4.1.2. An AAM texture synthesis yields $\boldsymbol{t}(\boldsymbol{h}_c)$ (see eq. 4.26).

Texture alignment The values of the warped texture and the synthesized texture are defined in completely different intervals. A synthesized texture is very close to a normalized vector, containing very small values, while a warped texture contains values between (0..255). Hence the texture alignment transformation A is necessary to project the values of the warped texture into the range of the synthesized texture. This is of great numerical importance. Otherwise the

4.3.1 Objective Function

warped texture would dominate the difference between warped and synthesized texture. Furthermore, the quality of the entire optimization would suffer from limited computational accuracy, due to the multi-digit mantissa.

Therefore the warped texture is aligned to the mean texture $\bar{t}$ by the texture alignment transformation A to project it into the AAM texture domain, resulting in the aligned texture t'. The texture alignment transformation A is defined as

$$A : t' = \left(t - \mathbf{1} \frac{t \cdot \mathbf{1}}{|t|} \right) \cdot \frac{1}{t \cdot \bar{t}} \quad (4.31)$$

Here · denotes the scalar product of two vectors and $\mathbf{1}$ a vector containing all ones. The texture t is aligned to the mean texture $\bar{t}$ by liberation from the mean and then scaling with $1/(t \cdot \bar{t})$.

Pose transformation As s is synthesized in the normalized PCA space using the combined model coefficient vector h_c, it must be transformed properly into the pixel coordinate space of the image that is to be analyzed applying the translation (t_x, t_y), the scaling s_r and the rotation θ to it (see fig. 4.6 for an illustration). Translations are applied to the face coordinate space so that a translation t_x is performed along the face x-axis and a translation of t_y along the face y-axis.

Initialization Usually AAMs require a good start estimation. The system for face analysis described in this work provides a good-natured initial estimation by automatically locating the head and the eyes within the face image 3. The initial coefficient vector $h_{[0]}$ is set to the zero vector which corresponds to an synthesis where $s(h_{c[0]}) = \bar{s}$ and $t(h_{c[0]}) = \bar{t}$ with zero translation, an actual scaling factor of 1 and no rotation. It is characteristic for the Active Appearance Model that a synthesis given a zero coefficient vector always results in the mean shape $\bar{s}$ and the mean texture $\bar{t}$. To preserve the linear nature of the model for the pose and texture coefficients likewise and get the identity transformation for a zero extended coefficient vector h it is necessary to define these coefficients specifically. The scaling s is defined as

$$s = s_r \cos \theta - 1 \quad (4.32)$$

and the rotation r is chosen to be

$$r = s_r \sin \theta \quad (4.33)$$

so that $t_x = 0$, $t_y = 0$, $s = 0$ and $r = 0$ result in the identity transformation with no translation, a scaling $s_r = 1$ and a rotation $\theta = 0$. s_r is the actual scaling applied to the shape and θ the actual rotation. It can be shown that

$$s_r = \sqrt{1 + 2s + s^2 + r^2} \quad (4.34)$$

$$\theta = \operatorname{sign}(r) \cdot \arccos\left(\frac{1+s}{s_r}\right) \quad (4.35)$$

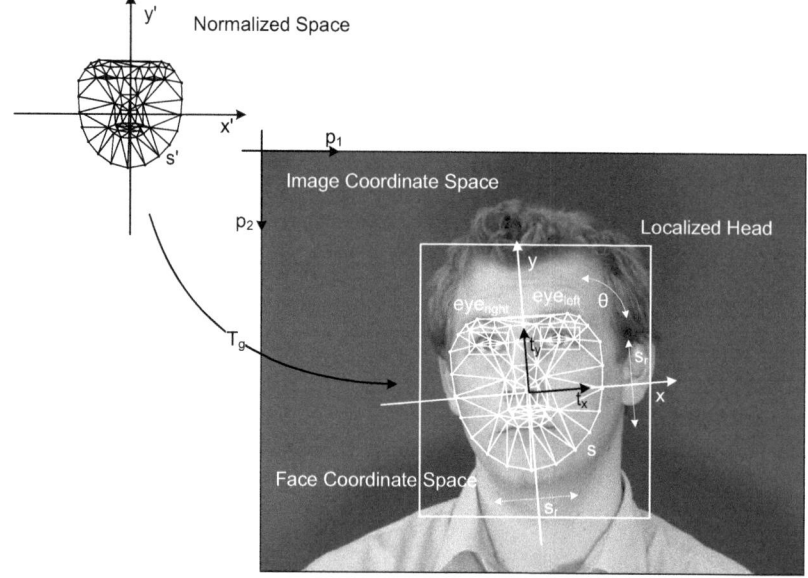

Figure 4.6: Transformations during AAM coefficient optimization

are solutions for the equations 4.32 and 4.33 solved for s_r and θ.

The same argumentation applies to the texture coefficients intensity m and brightness b. The adjusted texture t' is defined as

$$t' = (m+1)t + b \qquad (4.36)$$

As both, the texture as well as the shape, depend on the extended coefficient vector h, it is obvious that a minimal difference image $r(h)$ and hence a minimal error energy E(h) imply a locally "optimal" configuration of h.

4.3.2 Offline Prediction

This section describes an offline optimization approach where the gradient required for optimization is estimated a-priori during AM training [27]. This heavily reduces the amount of operations during the application of the AAM.

A first order Taylor expansion of $r(h)$ around h in $h + \delta h$ leads to

$$r(h + \delta h) = r(h) + J\delta h \qquad (4.37)$$

where $J = \frac{\partial r}{\partial h}$ is the Jacobian matrix of $r(h)$. The element in the ith row and jth column is defined as $J_{i,j} = \frac{\partial r_i}{\partial h_j}$.

To minimize the error energy E(h) from eq. 4.27 it is desired to determine a proper value for δh so that $\frac{1}{2}\|r(h + \delta h)\|^2 \overset{!}{=} \mathbf{0}$.

4.3.3 Numerical Estimation of the Jacobian Matrix

By setting $r(h+\delta h) = 0$, applying the pseudo inverse of J, and finally solving equation 4.37 for $r(h)$ the solution

$$\begin{aligned} -r(h) &= J\delta h \\ -J^T r(h) &= J^T J \delta h \\ \delta h &= -R r(h) \end{aligned} \qquad (4.38)$$

is obtained where $R = (J^T J)^{-1} J^T$. R can be pre-calculated once during training and used to predict the gradient during coefficient optimization.

The Jacobian matrix J can be numerically estimated by generating a number of sample faces and displacing each coefficient slightly from its calculated optimum. The weighted sum of all differences forms the estimated partial gradient for a specific coefficient (se section 4.3.3).

A major drawback of the offline prediction approach is that a good start estimation of the face on the original image is required because this method can even out only minor deviations from the optimum. The primary advantage is its low computational complexity. As such a-priori prediction it is a good compromise between speed and quality.

4.3.3 Numerical Estimation of the Jacobian Matrix

The Jacobian matrix J is calculated by building *left/right differences*. Each coefficient h_i of the coefficient vector h is displaced by a small amount for a number of d randomly generated sample faces. The results are summed up and smoothed by an appropriate normalized Gaussian kernel w. Thus one element of the *Jacobian matrix* in the ith row and jth column $J_{i,j} = \frac{\partial r_i}{\partial h_j}$ can be numerically approximated by

$$\frac{\partial r_i}{\partial h_j} = \sum_{k=0}^{d-1} w(\delta v_{kj})(r_i(h + \delta v_{kj} e_j) - r_i(h)) \qquad (4.39)$$

where e_j is a unit vector with a 1 at position j and all other elements set to zero. Therefore, the scalar δv_{kj} is only added to the jth coefficient of vector h. The value for δv_{kj} is a fixed or random displacement for the kth sample and jth coefficient.

Known sample faces can be generated by randomly initializing a set $\mathcal{V}$ of d coefficient vectors h_i, $0 \leqslant i < d$. These vectors will later be used to synthesize random faces with the model. Optimally the first six elements of these vectors, representing the pose and texture coefficients (see equation 4.29) are set to zero and the sub-vector h_c, representing the combined model coefficient vector is initialized with values up to three times the standard deviation $\sigma_i = \sqrt{\lambda_{c,i}}$ in the corresponding PCA dimension. In this case the substitution of 4.30 into 4.39 evaluates to

$$\frac{\partial r_i}{\partial h_j} = \sum_{k=0}^{d-1} w(\delta v_{kj}) \left(t_{orig}(h + \delta v_{kj} e_j) - t_{synth}(h + \delta v_{kj} e_j) \right) \qquad (4.40)$$

$$-t_{orig}(h) + t_{synth}(h))$$
$$= \sum_{k=0}^{d-1} w(\delta v_{kj}) (t_{orig}(h + \delta v_{kj} e_j) - t_{synth}(h + \delta v_{kj} e_j)) \quad (4.41)$$

because as we already use synthesized images for t_{orig} and obtain $t_{synth}(h) - t_{orig}(h) = 0$. See listing 4.1 for a pseudo code example of the estimation algorithm and figure 4.7 for an illustration. The first column of the figure shows two randomly generated examples used during the numerical estimation of the Jacobian matrix. In the first row the translation coefficients t_x is modified, in the second row the rotation coefficient r is varied. The last image in each row visualizes the numerical estimation of the gradient $\frac{\partial r}{\partial h_j}$ for the jth coefficient.

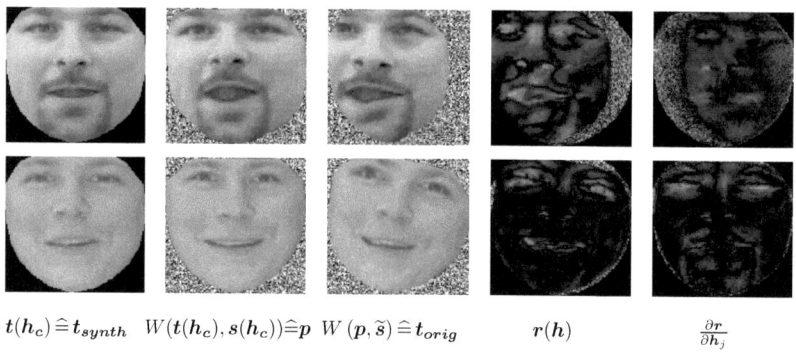

$t(h_c) \hat{=} t_{synth}$ $W(t(h_c), s(h_c)) \hat{=} p$ $W(p, \tilde{s}) \hat{=} t_{orig}$ $r(h)$ $\frac{\partial r}{\partial h_j}$

Figure 4.7: Numerical Estimation of the Jacobian matrix

From left to right the first image in each row shows randomly synthesized textures $t(h_c)$. These textures are warped into a synthesized shape $s(h_c)$ (second image). Herein, we introduced an modification of the standard AAM algorithm: The background of the images in column two and three shows pixels of white noise. Thus, during the analysis of a face image the optimization especially of the pose parameters becomes independent of the image background. The third image is perturbed because it has been warped from a slightly "mispositioned" source shape $\tilde{s}$ over the second image to the mean shape $\bar{s}$ resulting in textures t_{orig}. Here very large displacement values δv_{kj} have been chosen for a better visualization. The fourth image shows the difference between t_{orig} and t_{synth} (here $t_{synth} = t(h_c)$ because no model coefficients are modified, only the pose coefficients t_x and r). Eventually, the last image visualizes the estimated gradient $\frac{\partial r}{\partial h_j}$.

Listing 4.1: Jacobian matrix numerical estimation algorithm

```
1  // Set all elements of the Jacobian to zero
2  J = (0 | ... | 0)
3
4  // Initialize set V with random samples h_k ∈ R^v
```

4.3.4 Iterative Optimization

```
 5  V = {h_0,...,h_{d-1}})
 6
 7  for each h_k ∈ V do {
 8
 9      // Synthesize a texture and a shape
10      t = synthTexture(h_k)
11      s = synthShape(h_k)
12
13      // Warp the synthesized texture t from
14      // the mean shape s̄ to the shape s
15      t = warp(t, s̄, s)
16
17      // Loop over each coefficient
18      for j = 0 to v-1 do {
19
20          // Displace coefficient j by δv_{kj}
21          h'_k = h_k + v_{kj} e_j
22
23          // Synth. a texture and shape with displaced coeff.
24          s̃ = synthShape(h'_k)
25          t_synth = synthTexture(h'_k)
26
27          // Warp the synthesized texture t from the
28          // synthesized shape s̃ to the mean shape s̄
29          t_orig = warp(t, s̃, s̄)
30
31          // Calculate the weighted sum
32          J_{*,j} = J_{*,j} + w(v_{kj})(t_orig - t_synth)
33      }
34  }
```

4.3.4 Iterative Optimization

For a graphical illustration of the AAM coefficient optimization refer to figure 4.8. The optimization starts with an initial coefficient vector $\boldsymbol{h}_{[0]} = \boldsymbol{0}$ which is used to synthesize a shape $\boldsymbol{s}(\boldsymbol{h}_{c[n]})$ and a texture $\boldsymbol{t}(\boldsymbol{h}_{c[n]})$, where i is the index of the current iteration with $i \in \{0, \ldots, n-1\}$. During each iteration the synthesized shape is positioned on the input image $\boldsymbol{p}$, respecting the pose coefficients $t_x[n]$, $t_y[n]$, $s_{[n]}$, and $r_{[n]}$. The image region below the synthesized shape is warped to the mean shape $\bar{\boldsymbol{s}}$ producing a texture $\boldsymbol{t}$. The texture alignment transformation A aligns $\boldsymbol{t}$ to the mean texture $\bar{\boldsymbol{t}}$ to project it into the AAM texture synthesis domain and produces values that lie in the same interval like those of the synthesized texture $\boldsymbol{t}(\boldsymbol{h}_{c[n]})$. Additionally the texture adjustment coefficients $m_{[n]}$ and $b_{[n]}$ are applied to adapt the intensity and brightness of the warped texture to the intensity and brightness of the synthesized texture. Finally the difference $\boldsymbol{r}(\boldsymbol{h}_{[n]})$ between these two textures is optimized iteratively by applying the update rule from eq. 4.42. The algorithm terminates after a specific number of iterations or the error difference $\Delta E(\boldsymbol{h})$ between two consecutive iterations falls below a defined threshold.

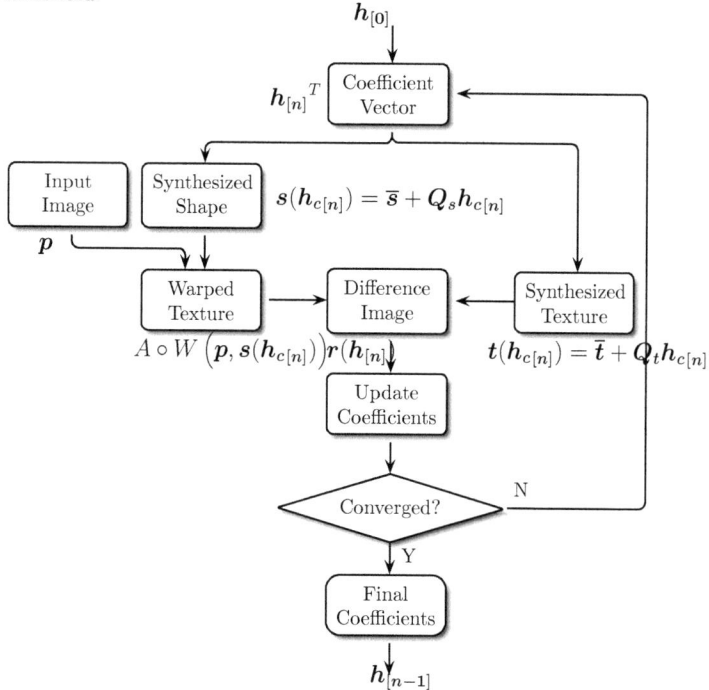

Figure 4.8: Schematic illustration of the AAM coefficient optimization algorithm

The update rule for $\boldsymbol{h}_{[n]}$ at iteration n can in general be written as

$$\boldsymbol{h}_{[n+1]} = \boldsymbol{h}_{[n]} + \alpha_{[n]} \delta \boldsymbol{h}_{[n]} \qquad (4.42)$$

The algorithm starts with an initial estimation $\boldsymbol{h}_{[0]}$ and a step width α. Thereby, $\delta \boldsymbol{h}_{[n]}$ can directly represent a difference vector of the current position in the search space $\boldsymbol{h}_{[n]}$ to the new position $\boldsymbol{h}_{[n+1]}$ where a minimization of the objective function is performed. In this case of *Sampling* of the search space is $\alpha_{[0]} = \alpha_{[n]} = 1$. In case of a *Gradient Descent* approach, $\delta \boldsymbol{h}_{[n]}$ constitutes a gradient consisting in a direction and a steepness. The previously described Offline Prediction, as it is proposed by the standard AAM algorithm, provides a predicted gradient. It is further suggested to update $\alpha_{[n]}$ between two consecutive iterations applying a decay term. One example multiplies α with a fixed factor $\beta < 1$ after each iteration so that $\alpha_{[n]} = \alpha \cdot \beta^n$. Another approach to update the step size $\alpha_{[n]}$ is setting $\alpha_{[n]} = \frac{\alpha}{(1+n\cdot\gamma)}$ with $\gamma > 0$. Our experiments have shown that the AAM shows generally better convergence with the first decay term and an initial step width $\alpha_{[0]} \approx 1.5$ and $\beta \approx 0.7$.

Chapter 5

Derivatives and Advancements of Active Appearance Models

5.1 A Survey on Active Appearance Models and Variants

AAMs have originally been introduced by Edwards et al. in [37], and have later been expanded by Cootes et al. in [21] and [19]. Since their introduction, AAMs have found many applications in a variety of areas where one desires to align, track, or interpret images of deformable objects, and various variations of the basic AAM algorithm have been developed. See [22] for an experimental comparison of some important AAM variations.

After the introduction of the original algorithm, Edwards et al. have introduced an extension to the basic algorithm to handle color images, and also provided an enhanced search algorithm that is more robust against occlusions [35]. In [25], Cootes et al. have shown that multiple AAMs can be used to model human faces from any view point and that these models can be used to track faces and estimate head poses. Cootes et al. have also shown that prior information, such as initial estimates about the locations of the eyes, can be used to constrain AAM search to obtain more reliable results [23]. In [24], Cootes et al. have demonstrated that using a linearly transformed representation of the edge structures at each pixel instead of the pixel values provides more accurate matching of the model to the target image and, in [20], they have proposed subsampling techniques to speed up AAM convergence for the expense of loosing some accuracy.

Baker et al. have proposed to use an inverse compositional approach to match the AAM to the target image, instead of the additive approach used in the basic AAM algorithm [6]. They consider an Appearance Model where the shape and texture coefficients are not combined into appearance coefficients and show that their approach provides increased efficiency during the matching procedure. Their formulation for the matching algorithm requires the warp functions to satisfy certain properties, which are not satisfied by the warps used in AAM. Therefore, they use first-order approximations of the inversion and composition operators.

AAMs have been extensively applied to medical image processing. Mitchell et al. have developed a multistage hybrid AAM to automatically segment left and right cardiac ventricles from magnetic resonance (MR) images [78]. Their hybrid approach combines AAMs with Active Shape Models (ASMs) to achieve more robustness against the risk of being trapped in local minima. Bosch et al. have extended AAMs to Active Appearance Motion Models (AAMMs) that enhance AAMs by including time dependent information. Their goal is to automatically segment echo cardiographic image sequences in a time continuous manner. Therefore, they consider the whole image sequence as a single shape-intensity pattern, and construct a single AAM that can describe both the appearance of the heart at a certain time and its dynamics throughout the cardiac cycle. This provides a time continuous segmentation, and eliminates the need for constructing different AAMs to segment the heart at different phases of the cardiac cycle. They also propose a nonlinear intensity normalization technique to deal with the non-Gaussian nature of the distribution of the intensity values in ultrasound images, which is shown to provide a significant improvement in performance. Mitchell et al. have proposed a three-dimensional AAM to segment volumetric cardiac MR images and echo cardiographic image sequences [79]. This 3-D extension of AAM provides a successful segmentation of 3-D images in a spatially and temporally consistent manner.

Motivated by the AAM algorithm, Hou et al. have developed Direct Appearance Models (DAMs) where they use the texture to directly predict the shape during the iterations of the parameter updates [52]. This approach no longer combines the shape and texture parameters into appearance parameters like AAM does. Li et al. have later extended this approach for multi-view face alignment by training multiple models for different poses of the human face [68]. In relation to his work, Yan et al. have developed texture-constrained ASMs (TC-ASMs), where the shape update predicted by an ASM is combined with the shape constraint provided by a global texture model like the one in AAM [126]. They show that such an approach performs better than ASM or AAM alone.

Stegmann and Larsen propose to generalize the concept of texture in AAM to include any measurement over the target image selected according to the particular class that is being modeled [106]. For the specific case of human faces, they propose a representation that includes the intensity value, the hue, and the edge strength and show that an AAM which uses this representation provides a better result than an AAM that works only on gray scale intensities. In [107], Stegmann et al. propose a few extensions to AAM that include an enhancement to the shape modeling, an initialization scheme to make the system fully automated, and a simulated annealing approach to fine tune the AAM parameters after the basic algorithm has converged. In [36], Edwards et al. use AAMs for face recognition, and in [5], Ahlberg uses AAMs to track faces and to automatically extract MPEG-4 facial animation parameters. In [105], Stegmann demonstrates that AAMs can be used for general object tracking.

Furthermore, Batur and Hayes proposed an adaptive AAM where they abandon the fixed gradient matrix approach of the basic AAM, and replace it with a linearly adaptive matrix that is updated according to the composition of the target texture [8]. Their approach starts with the observation that a fixed gradient matrix inevitably specializes to a certain region of the texture space and does not work well as the target image's texture moves away from this region. In general, the gradient matrix depends on both the shape and the texture of the target image. Since the gradient is computed in a normalized frame where the shape is normalized to the mean shape, one fixed matrix is a good estimate for the gradient matrices at different shapes. However, the same desirable property does not hold for different textures [9].

While our implementation of AM generation and the offline predicted gradient for AAM optimization already incorporates several of the proposed improvements mentioned above, we investigate novel techniques for the generation and the optimization process. The following sections introduce shape and texture models which base on the data modeling technique of Non-negative Matrix Factorization (section 5.2.1). Due to our computationally efficient implementation and the recent developments in processing power of modern CPUs and GPUs (see section 5.4) we applied online optimization methods to face analysis with AAMs (section 5.3). For a reliable comparison of the different approaches, we present evaluation measures following human perception (section 5.5).

5.2 Appearance Models based on Non-negative Matrix Factorization

The standard AAM algorithm described in chapter 4 strives to reduce the variance in the appearance of objects to the shape and texture. Thereby, the shape and texture variance is modeled by several Principal Component Analyses in a holistic way. However, for complex objects like human faces we can assume a wide independence in the appearance of separated object parts, i.e. the shape and texture of the eyes is independent from the mouth and vice versa. Therefore, a parts based representation of the training images would lead to a higher flexibility of the model in generation of a wider range of face instances.

The Non-negative Matrix Factorization (*abbrev.* 'NMF') has shown this behavior in the works of Lee and Seung [64] in 1999. However, localized data is a by-product of NMF and as shown in [34] it depends on properties of the data corpus, whether the result is spatially localized or not. Hence from various sites additional constraints were introduced which provide localness in NMF image analysis, e.g. Local NMF [117]. In LNMF the cost function is changed, in order to maximize the sparseness of the coefficient matrix H, the expressiveness, and the orthogonality of the basis Φ which spans the conical NMF vector space. Therefore, the update rules took suitable modifications. Furthermore, for implementation of the so called non-smooth NMF (*abbrev.* 'nsNMF') Hoyer [53] and Pascual-Montano et al. [90] tried to make both matrices W and H sparse by application of additional sparseness constraints.

Hence NMF and its variants found a broad range of applications in image processing tasks, such as face recognition [118], facial expression recognition with LNMF [13] and low-resolution brain electromagnetic tomography with nsNMF [90]. With the formulation of statistical shape and texture models based on the Non-negative Matrix Factorization, this work presents a fully novel advancement of the traditional AAM methodology.

The following section gives a comprehensive introduction in the NMF algorithm, since the common publications on Non-negative Matrix Factorization omit crucial steps in the mathematical argumentation and proof [64] [65] [34]. Further the formulation of the NMF-based generation of Appearance Models is given with the consequences for the NMF-AAM object analysis.

5.2.1 Data Modeling with Non-Negative Matrix Factorization

We seek a decomposition of the matrix $D \in \mathbb{R}^{c \times p}$ into the two matrices $W \in \mathbb{R}^{c \times r}$ and $H \in \mathbb{R}^{r \times p}$. The columns of matrix D consist of data vectors representing a data set to be transformed in a new r-dimensional, conical coordinate system. The data vectors may constitute the pixel arrays of (face-) images, AM-shape vectors, or frequency coefficients of digital sound files. The decomposition with the non-negative constraint reads as follows

$$D = WH + U \qquad D_{i,j}, W_{i,\mu}, H_{\mu,j} \geqslant 0 \qquad (5.1)$$

with $0 \leqslant i < c - 1$, $0 \leqslant j < p - 1$ and $0 \leqslant \mu < r - 1$ and U being the residual error. First we have to define a proper cost function, measuring the error between the original data D and the approximation WH, then we must find a way of minimizing this cost function in respect to W and H.

The NMF is actually a conical coordinate transformation. See figure 5.1 for a graphical interpretation. The two basis vectors w_1 and w_2 describe a cone which encloses the dataset $\mathcal{D}$. Due to the non-negative constraints only points within this cone can be re-constructed through linear combination of these basis vectors:

$$d' = (w_1, w_2) \cdot (h_1, h_2)^T \qquad (5.2)$$

The factorization of $D \approx WH$ is not necessarily unique. For example one can apply the following transformation using the arbitrary matrix A and its inverse A^{-1}:

$$WH = WAA^{-1}H \qquad (5.3)$$

If the two matrices $\hat{W} = WA$ and $\hat{H} = A^{-1}H$ are positive semi-definite then another factorization $D \approx \hat{W}\hat{H}$ exists. Such a transformation is always possible if A is an invertable non-negative monomial matrix. A matrix is called monomial if there is exactly one element different from zero in each row and column. If A is a non-negative monomial matrix the result of this tranformation is simply a scaling and permutation of the original matrices.

5.2.1 Data Modeling with Non-Negative Matrix Factorization

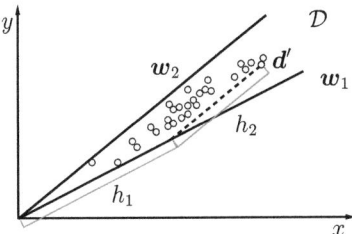

Figure 5.1: Graphical illustration of NMF: Non-negative matrix factorization as conical coordinate transformation

5.2.1.1 Cost Functions

Lee and Seung propose two different cost functions to be optimized [65]. The first is the square Euclidean distance

$$\underset{W,H}{\operatorname{argmin}} ||D - WH||^2 = \sum_{0 \leq i < c,\ 0 \leq j < p} (D_{i,j} - (WH)_{i,j})^2 \qquad (5.4)$$

the second is the Kullback-Leibler divergence

$$\underset{W,H}{\operatorname{argmin}} D_{KL}(D||WH) = \sum_{0 \leq i < c,\ 0 \leq j < p} D_{i,j} \log \frac{D_{i,j}}{(WH)_{i,j}} - D_{i,j} + (WH)_{i,j} \qquad (5.5)$$

that is also sometimes known as *information gain*, *relative entropy* or *information divergence* for $\sum_{i,j} D_{i,j} = \sum_{i,j} (WH)_{i,j} = 1$.

Both measures are suitable as cost functions because $||D-WH||^2, D_{KL}(D||WH) > 0$ for $D \neq WH$ and $||D - WH||^2, D_{KL}(D||WH) = 0$ for the optimal case $D = WH$. Within this report only the square Euclidean distance is discussed. For an in-depth study of the Kullback-Leibler divergence please refer to the work of Lee and Seung [65].

5.2.1.2 Proof of Convergence

Theorem 5.2.1 *It can be shown, that the square eculidian distance measure as proposed in section 5.2.1.1 is non-increasing under the following iterative update rules:*

$$H_{a\mu} \leftarrow H_{a\mu} \frac{(W^T D)_{a\mu}}{(W^T W H)_{a\mu}} \qquad W_{ia} \leftarrow W_{ia} \frac{(DH^T)_{ia}}{(WHH^T)_{ia}} \qquad (5.6)$$

for $0 \leq a < r$, $0 \leq \mu < p$ and $0 \leq i < c$.

These update rules can be interpreted as diagonally rescaled gradient descent. A standard gradient descent would require a learning rate α that is sufficiently small. The updates rules above in contrast correspond to a rather large learning rate. It is interesting that the algorithm still converges into a local minimum.

It can already be noted, that the quotient in both equations converges against one if $\boldsymbol{WH} \approx \boldsymbol{D}$. This means that a local minimum has been reached. In practice the algorithm terminates if the change of the respective matrix is sufficiently small.

The proof of convergence is quite similar to that of the Expectation Maximization (*abbrev.* 'EM') [30][123] algorithm. As the cost function F cannot be minimized directly, an auxiliary function G is defined that can be minimized analytically. The idea is that minimizing the auxiliary function will minimize the real cost function F. This is only shown for the first update rule that iteratively updates $\boldsymbol{H}$. The proof for $\boldsymbol{W}$ works exactly the same way and thus is not shown.

First the definition for auxiliary functions is introduced, next an auxiliary function for the square Euclidean distance measure is defined and proven an auxiliary function for this specific cost function. We finally minimize the auxiliary function to derive update rules for the minimization of the cost function.

Determination of an Auxiliary Function

Definition 5.2.1 *The function $G(\boldsymbol{h}, \boldsymbol{h}^{[t]})$ is an auxiliary function for $F(\boldsymbol{h})$ if*

(a) $G(\boldsymbol{h}, \boldsymbol{h}^{[t]}) \geqslant F(\boldsymbol{h})$ and

(b) $G(\boldsymbol{h}, \boldsymbol{h}) = F(\boldsymbol{h})$

With definition 5.2.1 the following theorem can be shown:

Theorem 5.2.2 *If $G(\boldsymbol{h}, \boldsymbol{h}^{[t]})$ is an auxiliary function for $F(\boldsymbol{h})$, minimizing $G(\boldsymbol{h}, \boldsymbol{h}^{[t]})$ in respect to $\boldsymbol{h}$ minimizes $F(\boldsymbol{h})$. The cost-function $F(\boldsymbol{h})$ is monotonically decreasing under the following update rule:*

$$\boldsymbol{h}^{[t+1]} = \underset{\boldsymbol{h}}{argmin}\, G(\boldsymbol{h}, \boldsymbol{h}^{[t]}) \tag{5.7}$$

Later the cost function F will be the square euclidean distance measure as introduced in equation 5.4 and G will be set to an appropriate auxiliary function that can be minimized analytically in respect to $\boldsymbol{h}$. We derive the update rules of equation 5.6 directly from equation 5.7.

Proof Per definition 5.2.1 it is $F(\boldsymbol{h}^{[t+1]}) \leqslant G(\boldsymbol{h}^{[t+1]}, \boldsymbol{h}^{[t]})$. According to equation 5.7, $\boldsymbol{h}^{[t+1]}$ minimizes $G(\boldsymbol{h}, \boldsymbol{h}^{[t]})$. Thus $G(\boldsymbol{h}^{[t+1]}, \boldsymbol{h}^{[t]})$ is less or equal than $G(\boldsymbol{h}^{[t]}, \boldsymbol{h}^{[t]})$, which equals $F(\boldsymbol{h}^{[t]})$ in line with definition 5.2.1. The following equation summarizes the steps above:

$$F(\boldsymbol{h}^{[t+1]}) \overset{def.}{\leqslant} G(\boldsymbol{h}^{[t+1]}, \boldsymbol{h}^{[t]}) \overset{min}{\leqslant} G(\boldsymbol{h}^{[t]}, \boldsymbol{h}^{[t]}) \overset{def.}{=} F(\boldsymbol{h}^{[t]}) \tag{5.8}$$

See figure 5.2 for an illustration. □

5.2.1 Data Modeling with Non-Negative Matrix Factorization

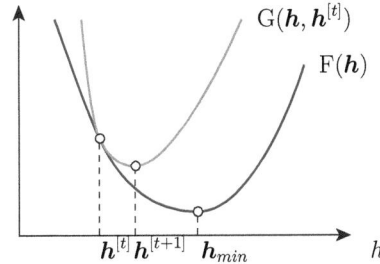

Figure 5.2: Illustration of theorem 5.2.2

Furthermore one can easily assure that there exists a sequence $h^{[0]} \cdots h_{min}$ so that

$$F(h^{[0]}) \geqslant \cdots \geqslant F(h^{[t]}) \geqslant F(h^{[t+1]}) \geqslant \cdots \geqslant F(h_{min}) \qquad (5.9)$$

converges into an approximated local minimum of $F(h)$.

To prove that the multiplicative update rules introduced in equation 5.6 converge into a local optimum, one must first define a proper auxiliary function G for the square Euclidean cost-function

$$F(h) = \frac{1}{2}||d - Wh||^2 \qquad (5.10)$$

Here the vector d represents an arbitrary vector from the matrix D and h is its corresponding encoding. By minimizing the auxiliary function G, update rules for the minimization of the cost function F can be derived according to equation 5.7. However, first it must be shown, that the defined auxiliary function fulfills the conditions of definition 5.2.1.

Lemma 5.2.3 *The function*

$$G(h, h^{[t]}) = F(h^{[t]}) + (h - h^{[t]})^T \nabla F(h^{[t]}) + \frac{1}{2}(h - h^{[t]})^T K(h^{[t]})(h - h^{[t]}) \qquad (5.11)$$

is an auxiliary function for the cost function

$$F(h) = \frac{1}{2}||d - Wh||^2 \qquad (5.12)$$

$$= \frac{1}{2} \sum_{0 \leqslant i < c} \left(d_i - \sum_{0 \leqslant a < r} W_{i,a} h_a\right)^2 \qquad (5.13)$$

if K is the diagonal matrix

$$K_{a,b}(h^{[t]}) = \delta_{a,b}(W^T W h^{[t]})_a / h_a^{[t]} \quad , \quad 0 \leqslant a, b < r \qquad (5.14)$$

with $\delta_{a,b}$ being the Kronecker-Delta function

$$\delta_{a,b} = \begin{cases} 1 & , a = b \\ 0 & , a \neq b \end{cases} \qquad (5.15)$$

Equation 5.11 is a slightly modified second order taylor expansion of the cost function F. Comparing equation 5.11 with the second order taylor expansion of F near the point $\boldsymbol{h}^{[t]}$:

$$F(\boldsymbol{h}) = F(\boldsymbol{h}^{[t]}) + (\boldsymbol{h} - \boldsymbol{h}^{[t]})^T \nabla F(\boldsymbol{h}^{[t]}) + \frac{1}{2}(\boldsymbol{h} - \boldsymbol{h}^{[t]})^T \boldsymbol{W}^T \boldsymbol{W}(\boldsymbol{h} - \boldsymbol{h}^{[t]}) \quad (5.16)$$

shows, that both eqations only differ in the last term.

Proof To show, that $G(\boldsymbol{h}, \boldsymbol{h}^{[t]})$ is an auxiliary function for $F(\boldsymbol{h})$, one must prove that according to definition 5.2.1 $G(\boldsymbol{h}, \boldsymbol{h}) = F(\boldsymbol{h})$ and $G(\boldsymbol{h}, \boldsymbol{h}^{[t]}) \geqslant F(\boldsymbol{h})$. The first case is rather trivial as

$$\begin{aligned} G(\boldsymbol{h}, \boldsymbol{h}) &= F(\boldsymbol{h}) + \underbrace{(\boldsymbol{h} - \boldsymbol{h})^T}_{=0} \nabla F(\boldsymbol{h}) + \frac{1}{2} \underbrace{(\boldsymbol{h} - \boldsymbol{h})^T}_{=0} \boldsymbol{K}(\boldsymbol{h}) \underbrace{(\boldsymbol{h} - \boldsymbol{h})}_{=0} \quad (5.17) \\ &= F(\boldsymbol{h}) \quad (5.18) \end{aligned}$$

In order to show that $G(\boldsymbol{h}, \boldsymbol{h}^{[t]}) \geqslant F(\boldsymbol{h})$, the auxiliary function $G(\boldsymbol{h}, \boldsymbol{h}^{[t]})$ is replaced by equation 5.11 and the cost function $F(\boldsymbol{h})$ by its second order taylor expansion as shown in equation 5.16. Both equations only differ in the last term so $G(\boldsymbol{h}, \boldsymbol{h}^{[t]}) \geqslant F(\boldsymbol{h})$ is equivalent to:

$$\begin{aligned} G(\boldsymbol{h}, \boldsymbol{h}^{[t]}) - F(\boldsymbol{h}) &= (\boldsymbol{h} - \boldsymbol{h}^{[t]})^T \boldsymbol{K}(\boldsymbol{h}^{[t]})(\boldsymbol{h} - \boldsymbol{h}^{[t]}) - (\boldsymbol{h} - \boldsymbol{h}^{[t]})^T \boldsymbol{W}^T \boldsymbol{W}(\boldsymbol{h} - \boldsymbol{h}^{[t]}) \quad (5.19)\\ &= (\boldsymbol{h} - \boldsymbol{h}^{[t]})^T (\boldsymbol{K}(\boldsymbol{h}^{[t]}) - \boldsymbol{W}^T \boldsymbol{W})(\boldsymbol{h} - \boldsymbol{h}^{[t]}) \geqslant 0 \quad (5.20) \end{aligned}$$

Now it must be shown, that equation 5.20 is greater or equal zero. This is the case, if the matrix $\boldsymbol{K}(\boldsymbol{h}^{[t]}) - \boldsymbol{W}^T \boldsymbol{W}$ is positive semi-definite, meaning that $\boldsymbol{K}(\boldsymbol{h}^{[t]}) - \boldsymbol{W}^T \boldsymbol{W} \geqslant 0$.

Definition 5.2.2 . *A matrix $\boldsymbol{M}$ is positive semi-definite, if it can be shown that $\boldsymbol{v}^T \boldsymbol{M} \boldsymbol{v} \geqslant 0$ for an arbitrary vector $\boldsymbol{v}$.*

To prove positive semi-definiteness, Lee and Seung [65] rescale $\boldsymbol{K}(\boldsymbol{h}^{[t]}) - \boldsymbol{W}^T \boldsymbol{W}$ with vector $\boldsymbol{h}$ and define the matrix $\boldsymbol{M}$ such that the element in the a-th row and b-th column can be written as follows:

$$\boldsymbol{M}_{a,b} = \boldsymbol{h}_a^{[t]} (\boldsymbol{K}(\boldsymbol{h}^{[t]}) - \boldsymbol{W}^T \boldsymbol{W})_{a,b} \boldsymbol{h}_b^{[t]} \quad (5.21)$$

Rescaling the matrix will later help to prove that it is positive semi-definite. This kind of rescaling is permitted as it does not violate the definition of positive semi-definiteness. We can easily show this by element-wise multiplication of the vector $\boldsymbol{v}$ from definition 5.2.2 and the vector $\boldsymbol{h}$ so that $(\boldsymbol{v} \odot \boldsymbol{h})^T \boldsymbol{M} (\boldsymbol{v} \odot \boldsymbol{h}) \geqslant 0$. This actually gives a new vector $\boldsymbol{v}' = \boldsymbol{v} \odot \boldsymbol{h}$ for which $\boldsymbol{v}'^T \boldsymbol{M} \boldsymbol{v}' \geqslant 0$ is also true. Moving the elements of $\boldsymbol{h}$ into the matrix $\boldsymbol{M}$ results in equation 5.21.

5.2.1 Data Modeling with Non-Negative Matrix Factorization

If it can be shown, that M is positive semi-definite, than $K(h^{[t]}) - W^T W$ is and thus equation 5.20 holds. According to definition 5.2.2 positive semi-definiteness of the matrix M can be verified as follows:

$$v^T M v = \sum_{0 \leqslant a,b < r} v_a M_{a,b} v_b \tag{5.22}$$

$$= \sum_{0 \leqslant a,b < r} v_a h_a^{[t]} (K(h^{[t]}) - W^T W)_{a,b} h_b^{[t]} v_b \tag{5.23}$$

$$= \sum_{0 \leqslant a,b < r} v_a h_a^{[t]} K_{a,b}(h^{[t]}) h_b^{[t]} v_b - v_a h_a^{[t]} (W^T W)_{a,b} h_b^{[t]} v_b \tag{5.24}$$

$$= \sum_{0 \leqslant a,b < r} v_a \underbrace{\delta_{a,b} (W^T W h^{[t]})_a h_b^{[t]} v_b}_{=0 \text{ for } a \neq b} - v_a h_a^{[t]} (W^T W)_{a,b} h_b^{[t]} v_b \tag{5.25}$$

$$= \sum_{0 \leqslant a,b < r} v_a^2 h_a^{[t]} \underbrace{(W^T W h^{[t]})_a}_{=\sum_{0 \leqslant b < r}(W^T W)_{a,b} h_b^{[t]}} - v_a h_a^{[t]} (W^T W)_{a,b} h_b^{[t]} v_b \tag{5.26}$$

$$= \sum_{0 \leqslant a,b < r} h_a^{[t]} h_b^{[t]} (W^T W)_{a,b} \cdot (v_a^2 - v_a v_b) \tag{5.27}$$

$$= \sum_{0 \leqslant a,b < r} h_a^{[t]} h_b^{[t]} (W^T W)_{a,b} \cdot (\frac{1}{2} v_a^2 + \frac{1}{2} v_b^2 - v_a v_b) \tag{5.28}$$

$$= \frac{1}{2} \sum_{0 \leqslant a,b < r} h_a^{[t]} h_b^{[t]} (W^T W)_{a,b} \cdot (v_a - v_b)^2 \tag{5.29}$$

$$\geqslant 0 \tag{5.30}$$

For explanation of some of the above transformations: From equation 5.24 to 5.25, $K(h^{[t]})$ is substituted and $h_a^{[t]}$ can be canceled in the first summand. Due to the Kronecker-Delta function the first summand is always zero for $a \neq b$. Therefore from equation 5.25 to equation 5.26 each occurrence of the index b can be replaced by a within the first summand. Next the expression $(W^T W h^{[t]})_a$ in equation 5.26 is rewritten as sum $(W^T W h^{[t]})_a = \sum_{0 \leqslant b < r} (W^T W)_{a,b} h_b^{[t]}$. As the parameter b runs over the same interval this sum can be merged with the outer sum. Furthermore $h_a^{[t]} h_b^{[t]} (W^T W)_{a,b}$ is factored out resulting in equation 5.27. Let $0 \leqslant a,b < r$. Generally v_a^2 can be written as $v_a^2 = \frac{1}{2} v_a^2 + \frac{1}{2} v_a^2$. For $a = b$ this is equivalent to $v_a^2 = \frac{1}{2} v_a^2 + \frac{1}{2} v_b^2$. For $a \neq b$ there will always exist two cases for which $h_a^{[t]} h_b^{[t]} (W^T W)_{a,b} = h_b^{[t]} h_a^{[t]} (W^T W)_{b,a}$ due to the symmetry of $W^T W$:

$$\cdots + h_a^{[t]} h_b^{[t]} (W^T W)_{a,b} \cdot (\frac{1}{2} v_a^2 + \frac{1}{2} v_a^2 - v_a v_b) + \cdots + h_b^{[t]} h_a^{[t]} (W^T W)_{b,a} \cdot (\frac{1}{2} v_b^2 + \frac{1}{2} v_b^2 - v_b v_a) + \cdots \tag{5.31}$$

This can be subsumed to

$$\cdots + h_a^{[t]} h_b^{[t]} (W^T W)_{a,b} \cdot (\frac{1}{2} v_a^2 + \frac{1}{2} v_b^2 - v_a v_b + \frac{1}{2} v_b^2 + \frac{1}{2} v_a^2 - v_b v_a) + \cdots \tag{5.32}$$

and summands can be exchanged so that

$$\cdots + h_a^{[t]} h_b^{[t]} (W^T W)_{a,b} \cdot (\frac{1}{2} v_a^2 + \frac{1}{2} v_b^2 - v_a v_b) + \cdots + h_b^{[t]} h_a^{[t]} (W^T W)_{b,a} \cdot (\frac{1}{2} v_b^2 + \frac{1}{2} v_a^2 - v_b v_a) + \cdots \tag{5.33}$$

Re-sorting the summands in the described manner leads to equation 5.28. Now it is possible to apply the second binomial formula which finally results in equation 5.29.

As $h_a^{[t]} \geqslant 0$, $(\boldsymbol{W}^T\boldsymbol{W})_{a,b} \geqslant 0$ and $(v_a - v_b)^2 \geqslant 0$ for $0 \leqslant a, b < r$, this proves positive semi-definiteness for $\boldsymbol{M}$. Therefore $\boldsymbol{K}(\boldsymbol{h}^{[t]}) - \boldsymbol{W}^T\boldsymbol{W}$ is positive semi-definite and thus is $(\boldsymbol{h} - \boldsymbol{h}^{[t]})^T(\boldsymbol{K}(\boldsymbol{h}^{[t]}) - \boldsymbol{W}^T\boldsymbol{W})(\boldsymbol{h} - \boldsymbol{h}^{[t]}) \geqslant 0$. As this is equivalent to $\mathrm{G}(\boldsymbol{h}, \boldsymbol{h}^{[t]}) \geqslant \mathrm{F}(\boldsymbol{h})$ as shown before in equation 5.20 and with the additional knowledge, that $\mathrm{G}(\boldsymbol{h}, \boldsymbol{h}) = \mathrm{F}(\boldsymbol{h})$ as denoted by equation 5.18, the function $\mathrm{G}(\boldsymbol{h}, \boldsymbol{h}^{[t]})$ is an auxiliary function for the cost function $\mathrm{F}(\boldsymbol{h})$.

□

Minimization the Auxiliary Function As proposed in theorem 5.2.2, minimizing $\mathrm{G}(\boldsymbol{h}, \boldsymbol{h}^{[t]})$ in respect to $\boldsymbol{h}$ will lead to an update rule for $\boldsymbol{h}^{[t+1]}$. Setting $\mathrm{G}(\boldsymbol{h}, \boldsymbol{h}^{[t]})$ to

$$\mathrm{G}(\boldsymbol{h}, \boldsymbol{h}^{[t]}) = \mathrm{F}(\boldsymbol{h}^{[t]}) + (\boldsymbol{h} - \boldsymbol{h}^{[t]})^T \nabla \mathrm{F}(\boldsymbol{h}^{[t]}) + \frac{1}{2}(\boldsymbol{h} - \boldsymbol{h}^{[t]})^T \boldsymbol{K}(\boldsymbol{h}^{[t]})(\boldsymbol{h} - \boldsymbol{h}^{[t]}) \quad (5.34)$$

for which it has been shown in lemma 5.2.3 that it is a proper auxiliary function for $\mathrm{F}(\boldsymbol{h})$ and minimizing it by setting its gradient to zero finally produces the update rules that are sought for:

$$\frac{\partial \mathrm{G}(\boldsymbol{h}, \boldsymbol{h}^{[t]})}{\partial \boldsymbol{h}} = \nabla \mathrm{F}(\boldsymbol{h}^{[t]}) + \boldsymbol{K}(\boldsymbol{h}^{[t]})(\boldsymbol{h} - \boldsymbol{h}^{[t]}) \stackrel{!}{=} 0 \quad (5.35)$$

Solving equation 5.35 for $\boldsymbol{h}$ results in

$$\nabla \mathrm{F}(\boldsymbol{h}^{[t]}) + \boldsymbol{K}(\boldsymbol{h}^{[t]})(\boldsymbol{h} - \boldsymbol{h}^{[t]}) = 0 \quad (5.36)$$
$$\boldsymbol{K}(\boldsymbol{h}^{[t]})(\boldsymbol{h} - \boldsymbol{h}^{[t]}) = -\nabla \mathrm{F}(\boldsymbol{h}^{[t]}) \quad (5.37)$$
$$\boldsymbol{h} = \boldsymbol{h}^{[t]} - \boldsymbol{K}(\boldsymbol{h}^{[t]})^{-1} \nabla \mathrm{F}(\boldsymbol{h}^{[t]}) \quad (5.38)$$

According to equation 5.7 this is equivalent to

$$\boldsymbol{h}^{[t+1]} = \underset{\boldsymbol{h}}{\operatorname{argmin}}\, \mathrm{G}(\boldsymbol{h}, \boldsymbol{h}^{[t]}) \quad (5.39)$$
$$= \boldsymbol{h}^{[t]} - \boldsymbol{K}(\boldsymbol{h}^{[t]})^{-1} \nabla \mathrm{F}(\boldsymbol{h}^{[t]}) \quad (5.40)$$

Derivation of the Update Rule Now equation 5.40 will be rewritten. For this, the inverse of $\boldsymbol{K}(\boldsymbol{h}^{[t]})$ is required. As $\boldsymbol{K}$ is a diagonal matrix the inverse $\boldsymbol{K}^{-1}$ takes the form:

$$\boldsymbol{K}_{a,b}^{-1}(\boldsymbol{h}^{[t]}) = \delta_{a,b} \frac{h_a^{[t]}}{(\boldsymbol{W}^T \boldsymbol{W} \boldsymbol{h}^{[t]})_a} \quad (5.41)$$

5.2.1 Data Modeling with Non-Negative Matrix Factorization 61

Furthermore we need the partial derivatives $\frac{\partial F}{\partial h_a^{[t]}}$ of the function $F(h_a^{[t]})$:

$$\begin{align}
\frac{\partial F}{\partial h_a^{[t]}} &= -\sum_{0 \leq i < c}\left(d_i - \sum_{0 \leq j < r} W_{i,j} h_j^{[t]}\right) W_{i,a} \tag{5.42}\\
&= -(d - Wh^{[t]})^T w_a \tag{5.43}\\
&= -(w_a^T d - w_a^T Wh^{[t]}) \tag{5.44}\\
&= (W^T Wh^{[t]})_a - (W^T d)_a \tag{5.45}
\end{align}$$

with w_a being the a-th column of matrix W. Replacing $K^{-1}(h^{[t]})$ in equation 5.40 with equation 5.41 and writing the a-th element of vector $h^{[t]}$ explicitly results in:

$$h_a^{[t+1]} = h_a^{[t]} - \delta_{a,b} \frac{h_a^{[t]}}{(W^T Wh^{[t]})_a} \nabla F(h^{[t]}) \tag{5.46}$$

Due to the Kronecker-Delta function $\delta_{a,b}$ only the a-th partial derivative of $\nabla_a F(h^{[t]}) = \frac{\partial F}{\partial h_a^{[t]}}$ is selected so by replacing $\nabla_a F(h^{[t]})$ accordingly with the right hand side of equation 5.42, equation 5.46 can be transformed into:

$$\begin{align}
h_a^{[t+1]} &= h_a^{[t]} - \frac{h_a^{[t]}}{(W^T Wh^{[t]})_a} \frac{\partial F}{\partial h_a^{[t]}} \tag{5.47}\\
&= h_a^{[t]}\left(1 - \frac{(W^T Wh^{[t]})_a - (W^T d)_a}{(W^T Wh^{[t]})_a}\right) \tag{5.48}\\
&= h_a^{[t]} \frac{(W^T d)_a}{(W^T Wh^{[t]})_a} \tag{5.49}
\end{align}$$

Equation 5.49 represents the update rule for the a-th element of one column of matrix H. As each column of matrix H can be handled independently equation 5.6 holds and thus theorem 5.2.1 has been shown for H. By reversing the roles of W and H theorem 5.2.1 can also be proven for the matrix W. □

5.2.1.3 Algorithm Pseudocode

See listing 5.1 for the pseudocode of the NMF algorithm. As one can see, the implementation is fairly simple. The algorithm gets the raw non-negative data matrix D and the reduced r as input parameters and iteratively updates the matrices W and H according to equation 5.6. The algorithm assumes convergence, if the changes of W and H are sufficiently small. The method preprocess() in line 5 scales the values of each data vector to the range $[0;1]$. This methodology turned out to produce good results in practice. Typically r is chosen so that $r < \frac{nm}{n+m}$.

Our C/C++ implementation exploits the BLAS/ATLAS [10] [119] hardware optimizations as the NMF algorithm primarily requires linear algebra operations.

Listing 5.1: Pseudocode of the NMF Algorithm

```
function NMF(in data matrix D ∈ ℝ^(c×p), in reduced dimension r)
  returns base images W ∈ ℝ^(c×r), encoding H ∈ ℝ^(r×p) {

  W, H ← randomly initialize with values w, h ∈ [0..1]
  D ← preprocess(D);

  while (not converged) {
    WH ← calculate matrix-matrix product once
    // Update elements of H
    for (a = 0; a < r; a++) {
      for (μ = 0; μ < p; μ++) {
        H_{aμ} ← H_{aμ} (W^T D)_{aμ} / (W^T W H)_{aμ}
      }
    }
    // Update elements of W
    for (i = 0; i < c; i++) {
      for (a = 0; a < r; a++) {
        W_{ia} ← W_{ia} (D H^T)_{ia} / (W H H^T)_{ia}
      }
    }
  }
}
```

5.2.1.4 Experiments

Within this section we will present some experimental results on a qualitative basis. These results compare the capabilities of NMF and PCA. Note, that some images are scaled into the visible range for visualization purpose and may not resemble the original results exactly. Still, the same scaling factor is used for both methods so the images are still comparable.

Datasets NMF and PCA are applied to different datasets. Our MMER AAM Toolbox [32] provides the PCA implementation. The first dataset has mainly been created to illustrate the partial based approach of the NMF. It consists of six abstract square images, the first four (5.3(a) through 5.3(d)) show one white half-circle on black ground, differently oriented. The last two (5.3(e), 5.3(f)) can theoretically be composed from two of the four previous images respectively. See figure 5.3 for an illustration of the dataset D_{circ}.

The second dataset D_{face} has been created from the IMM face database [85]. This database consists of about 240 annotated color face images. These images have been normalized in shape using the MMER AAM Toolbox [32] and cropped so only the main face region is visible. The shape-normalization is an important

5.2.1 Data Modeling with Non-Negative Matrix Factorization

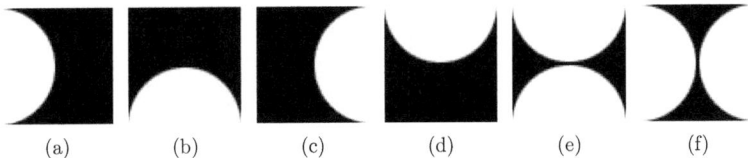

Figure 5.3: Abstract dataset D_{circ}

pre-processing step to make sure that each facial feature is located exactly in the same position in the pixel domain. Some example images are shown in figure 5.4. These images are the result of shape normalization and cropping. Visual distortion results from shape normalization.

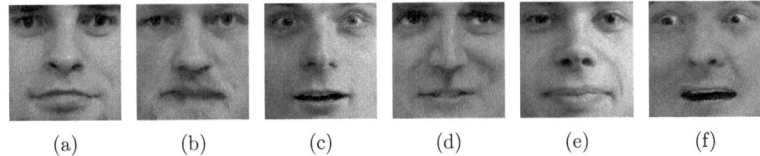

Figure 5.4: Face dataset D_{face}

Base Images We apply a PCA as well as the NMF algorithm to both datasets. For the first dataset D_{circ} the reduced r was set to $r = 4$ and also incorporated only the first four principal axes in case of the PCA. The assumption was that the NMF algorithm would most likely discover the inherent characteristics of the dataset and outperform the PCA in terms of re-construction quality. The result can be seen in figures 5.5 for the PCA and 5.6 for the NMF.

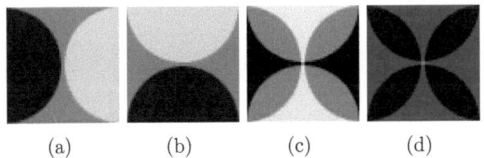

Figure 5.5: Base images of dataset D_{circ} after applying the PCA

The NMF obviously discovered, that the last two images of the dataset D_{circ} can be constructed from simpler parts and thus only the four most integral half circles are present in the base images. This allows a pretty intuitive re-construction. The results of the PCA are less intuitive and hard to interpret as PCA allows subtractive combination of base images as well.

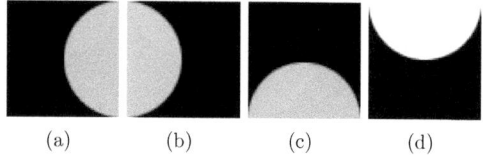

Figure 5.6: Base images of dataset D_{circ} after applying the NMF

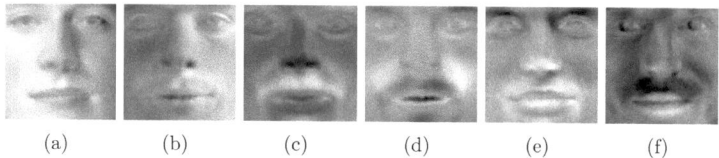

Figure 5.7: Base images of dataset D_{face} after applying the PCA

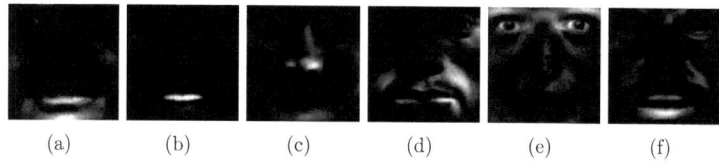

Figure 5.8: Base images of dataset D_{face} after applying the NMF

For the second dataset D_{face} the reduced r was set to $r = 30$ and incorporated only 30 principal axes respectively. The figures 5.7 and 5.8 show the computed base images for PCA and NMF.

Again the holistic approach of the PCA generates a global, distributed representation of the datasets principal characteristics. However, the NMF succeeds in finding local structures like eyes, nose, mouth etc.

Reconstruction The reconstructed images are illustrated in figures 5.9 and 5.10. The NMF based reconstructions are obviously better than those of the PCA. The reason for this is most likely that the PCA implements a holistic, rather global approach where the NMF gains its advantage from identifying local structures. Where the PCA fails to re-construct the original images without a noticeable error the NMF prevails.

For the second dataset both methods, PCA and NMF, succeed to reconstruct the original images but the result of the NMF is slightly sharper. For a direct comparison refer to figure 5.11. Notice the area around the eyes and the skin. The NMF produces a far more detailed result.

5.2.1 Data Modeling with Non-Negative Matrix Factorization

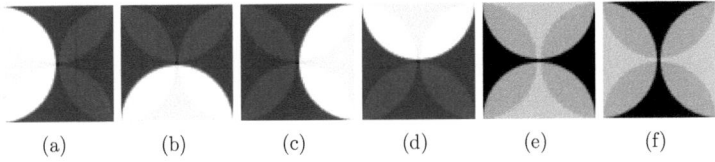

(a) (b) (c) (d) (e) (f)

Figure 5.9: Reconstructed images (PCA)

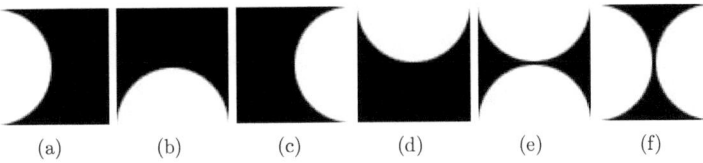

(a) (b) (c) (d) (e) (f)

Figure 5.10: Reconstructed images (NMF)

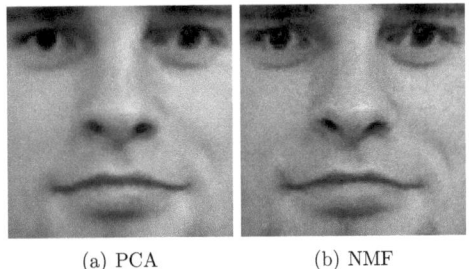

(a) PCA   (b) NMF

Figure 5.11: Face reconstruction

5.2.1.5 Conclusion

PCA and NMF are studied and compared to each other in a qualitative manner, the NMF algorithm is discussed in detail. Two datasets have been created to compare both algorithms and point out similarities as well as differences. Both methods provide good information reduction quality where the PCA follows a holistic and the NMF a part-based approach. This has advantages and disadvantages. PCA on the one hand can be calculated pretty fast with a definite result, where NMF is based on a diagonally rescaled gradient descent that suffers from well known problems that are typical for high dimensional optimization problems (local minima vs. global minima). The implementation is straightforward but computationally expensive. PCA is an orthogonal coordinate transformation, NMF finds a conical representation for the dataset. The drawback of this conical representation, that is rooted in the non-negative constraints, is the disability to construct data points that lie outside the cone. However, these theoretical limitations are practically negligible if one assumes that only a small range of variation in a specific domain has to be modeled. Depending on the quality and variance of the dataset itself, NMF will not always find an optimal part-based representation [34]. Recently methods came up, such as the Non-smooth Non-negative Matrix Factorization (nsNMF) [91], or the NMF with sparseness constraints [54], that tend to find local structures even better.

Possible Applications The field of possible applications ranges from general knowledge discovery and data mining tasks over computer vision and object recognition (e.g., face recognition) to bio-medical imaging. Both PCA and NMF are sophisticated yet computationally intense methods that reduce dimensionality without losing too much information. Both method allow to build a statistical model from a given training set e.g., facial images. This model is parameterizable through the encoding h which can be interpreted as parameter vector. Through automatic optimization the parameter vector h is chosen 'optimally' so that a synthesized image best resembles a given real image. The parameter vector h can be conceived as the low-dimensional representation of the rather high-dimensional face. It can be used for further classification (age, gender, ethnic group, emotion, etc.). Cootes et al. describe an interesting approach concerning deformable models of shape and texture [27].

NMF is not only suitable for computer vision tasks. It has already been used for semantic analysis [64] and can by applied to other data such as audio signals.

Outlook A major drawback of PCA based face recognition is its instability in the case of partial occlusion. One reason for this is the distributed, holistic representation of the dataset. Fully deformable NMF based models for face recognition should theoretically be relatively insensitive to partial occlusion.

5.2.2 Generation of Appearance Models with NMF

Starting from the standard PCA-AAM algorithm formulated in chapter 4, the Principal Component Analysis is replaced by the Non-negative Matrix Factorization in each step of the model generation process.

5.2.2.1 Shape Model

Analog to section 4.2.1 we apply the statistical analysis and information reduction on the matrix $S \in \mathbb{R}^{c \times p}$ of all aligned and normalized shape vectors in the training set. In case of NMF this means a factorization of S into the shape basis $\Phi_s \in \mathbb{R}^{L \times \mu_s}$ and the matrix of coefficient vectors $H_s \in \mathbb{R}^{\mu_s \times p}$.

$$S \approx \Phi_s H_s \qquad (5.50)$$

However, S is required to contain only positive values or zeros. This is ensured by adding the minimum of all x and the minimum of all y coordinates as offset to each landmark position prior to the factorization. This corresponds to a translation of all landmarks into the first quadrant.

5.2.2 Generation of Appearance Models with NMF

After an NMF analysis of S a new shape can be synthesized analog to eq. 4.16:

$$s = \Phi_s h_s \qquad (5.51)$$

Unlike the PCA variant, the new coordinate space Φ_s is not mean-free. Nevertheless, for the generation of the shape-free textures of the matrix T and during the AAM coefficient optimization, the mean shape $\bar{s}$ is required. It can be generated by application of the coefficient vector $\overline{h_s}$. For its determination the average of each line in H_s is computed.

$$\overline{h_s} = \frac{1}{p} H_s \cdot 1 \qquad (5.52)$$

$1 \in \mathbb{R}^p$ is a vector containing all ones.

Furthermore, the specific computation of the representation of all training shape vectors s_i in the new space can be skipped, since the NMF analysis already provides the coefficient vector h_{si} of shape s_i in the ith column of H_s.

5.2.2.2 Texture Model

Again, the AM generation provides for the triangulation of the mean shape and the warping of all training textures in the shape-free space. Since the matrix of shape-free textures T contains (color-)intensity values in the range of $[0; 255]$ the non-negativity requirement is naturally met. The application of NMF on T leads to a decomposition

$$T \approx \Phi_t H_t \qquad (5.53)$$

with the following matrix dimensionalities: $\Phi_t \in \mathbb{R}^{n \times \mu_t}$, $H_t \in \mathbb{R}^{\mu_t \times p}$.

The representation of the training textures in the new basis allows for the synthesis of a texture by evaluating $t = \Phi_t h_t$.

5.2.2.3 Combined Model

The combination of shape and texture to a combined model reads simpler than eq. 4.22, since the NMF bases are not mean-free.

$$C = \begin{bmatrix} KH_s \\ H_t \end{bmatrix} \qquad (5.54)$$

K is equal to the weight matrix in eq. 4.20. The combined matrix is again subject to an NMF decomposition:

$$C \approx \Phi_c H_c \qquad (5.55)$$

Further, the upper μ_s lines of Φ_c pertain to the shape model, the lower μ_t lines to the texture model. Thus Φ_c can be written as

$$\Phi_c = \begin{bmatrix} \Phi_{cs} \\ \Phi_{ct} \end{bmatrix} \qquad (5.56)$$

We can therefore express a shape s and a texture t as function of h_c as follows:

$$s = Q_s h_c \quad , \quad Q_s = \Phi_s K^{-1} \Phi_{cs} \tag{5.57}$$
$$t = Q_t h_c \quad , \quad Q_t = \Phi_t \Phi_{ct} \tag{5.58}$$

Finally, with the equations 5.50 to 5.57 the generation of an AM based on the Non-negative Matrix Factorization is fully specified. Although we find a unified notation for the description of both AAM variants, the actual differences in representation has several consequences as described in the following section.

5.2.2.4 Algorithmic Differences of PCA- and NMF-AAMs

Similar to the PCA modeling, the data reduction parameters μ_s, μ_t, and μ_c have high impact on the quality of the NMF result as described in 5.2.1. Unlike PCA, the ideal values have to be determined by running the NMF with different settings and evaluating the result. Objective quality measures are e.g. the residual error $U = D - \Phi H$ (compare eq. 5.1) or the target application in terms of pattern recognition results. The data reduction parameters of an PCA Appearance Model can be adjusted by the rate of the explained data variance to the *absolute* data variance after the PCA. The explained data variance can be measured by equation 5.61. Unfortunately, it is hard to determine the absolute data variance especially for the generation of the combined NMF model. For a NMF-AAM, e.g. built on a set of 50 images with distinctive facial expressions, we suggest to set the reduction parameter values in the range of $\mu_s \approx 28$, $\mu_t \approx 36$, and $\mu_c \approx 30$.

For initialization of the AAM coefficient optimization, it appears to be reasonable [27] to start with the mean shape $\bar{s}$ and the mean texture $\bar{t}$. In case of the PCA-AAM this translates into h_c being the zero vector due to the mean-free property of PCA. However, for NMF the values of $\overline{h_c}$ describing the mean shape and texture via equations 5.57 and 5.58 have to be computed [51]. This can be performed on basis of the matrix H_c of the combined model coefficients:

$$\overline{h_c} = \frac{1}{p} H_c \cdot 1 \tag{5.59}$$

Furthermore, during the optimization the search space for the ith coefficient $h_{c,i}$ should be limited to the range

$$\left[\overline{h_{c,i}} - 3\sqrt{\sigma_{c,i}^2} \; ; \; \overline{h_{c,i}} + 3\sqrt{\sigma_{c,i}^2}\right] \tag{5.60}$$

where $\sigma_{c,i}^2$ denotes the variance of the ith coefficient over the training set:

$$\sigma_{c,i}^2 = \frac{1}{p} \sum_{j=0}^{p-1} \left(H_{c i,j} - \overline{h_{c,i}}\right)^2 \tag{5.61}$$

While the Eigenvalue $\lambda_{c,i}$ of the PCA analysis already matches with the variance $\sigma_{c,i}^2$ of the ith basis vector, this has to be additionally computed from H_c for NMF-AAMs. On the one hand a specific computation of the mean and variance is necessary for NMF-AAMs, on the other hand they instantly provide the coefficient matrices H_s, H_t, and H_c, whereas those have to be determined explicitly in the PCA case.

A further difference exists in the way of the implementation of the texture model. For an Appearance Model of a texture resolution of 128×128 pixels, the dimensionality of T is approximately $40 \cdot 10^3 \times p$. The PCA on T requires a detour mentioned in 4.2.2. Thus, an unbearable computational effort of the Eigenvalue Decomposition of a covariance matrix in $\mathbb{R}^{40,000 \times n}$ is avoided whereas, the NMF can directly operate on T. However, this still constitutes the most computationally expensive step in AM generation.

The constraints of the NMF regarding the non-negativeness of the shape, texture, and combined matrices are either naturally met (T) or can be assured by transformation of all normalized shapes into the first quadrant in the 2D landmark space. As explained in section 5.2.1 the data representation in the NMF space bases on additive linear combination of the basis vectors. This would prohibit negative values in the coefficient vector h_c. However, the subsequent optimization algorithms assume the mean shape and texture to be in the origin of the optimization space. In order to preserve the compatibility of the NMF model with the implementation of the coefficient optimization, in each iteration n the mean coefficient vector $\overline{h_c}$ is added to $h_{c[n]}$ obtained from the optimization algorithm (see section 4.3). Likewise the mean shape $\overline{s}$ (eq. 5.52) must be subtracted from the generated shape of equation 5.57 before the warping transformation W can be correctly applied on the original texture of the analyzed image (see section 4.3.1). Therefore, with the mentioned compatibility requirements of the NMF-AAM implementation the equations 5.57 and 5.58 now read

$$s = Q_s \left(h_c + \overline{h_c} \right) - \overline{s} \qquad (5.62)$$

$$t = Q_t \left(h_c + \overline{h_c} \right) \qquad (5.63)$$

Although the NMF based Appearance Model is described by the same parameters $(Q_s, Q_t, \overline{s}, \overline{t})$ with the same semantic content, the algorithmic parameters of the optimization strategies require an extensive adaptation to the NMF-AM. This is mainly caused by the uniformer distribution of the explained variance over the basis vectors in the NMF space and the different codomain of the model coefficients. Details are given in the work of [51].

5.2.3 Conclusion

In this section we were able to retrace the relevant steps in the mathematical argumentation provided in commonly available publications on the Non-negative Matrix Factorization (*abbrev.* 'NMF') [65] [64]. Based on this theoretical understanding, exemplary experiments confirmed the reported behavior of NMF applied on image data [34], where a NMF decomposition of a set of images showing objects of a specific class tends to provide a localized and parts based representation. This property allows for an independent modeling and synthesis of parts of complex objects like a face with relatively independent eyes, nose, and mouse variants. Thus, we developed and formulated the model generation process of

Appearance Models based on NMF instead of Principal Component Analysis for the first time. Thereby, especially the non-negative constraints of NMF entail several differences in the preprocessing and representation of the shape and texture data. Despite the differences, a NMF-AM was formulated fully compatible with the subsequent coefficient optimization routines which are originally designed for PCA based AMs.

The presented experiments as well as the evaluation results (see chapter 6) show the capability of the NMF for a more precise representation of the training data at a greater dimensionality reduction compared to PCA. The limitation of the required dimensions is a great advantage, since it heavily decreases the computational effort of basically all coefficient optimization strategies. Unfortunately, the PCA still delivers a statistical data model with enhanced generalization abilities. Although this was not explicitly evaluated in this work, we suggest to apply NMF based AAMs for the more precise and more compact, hence computationally cheaper, analysis of objects with a low variance, such as person-specific AAMs.

5.3 Online Optimization of AAM Coefficients

Several experiments [46] have shown, that the offline gradient prediction scheme of the standard AAM algorithm described in section 4.3.2 fails to optimize Active Appearance Models under adverse circumstances. The discussed approach shows sensitivity to unbalanced or insufficient illumination conditions with bad brightness and contrast. In general, differences between the training samples for the gradient prediction and the analyzed images are problematic and lead to unsatisfactory results of the AAM re-synthesis and parametrization. Above all, the gradient used during optimization is the estimated mean of several gradients calculated during the training phase (refer to section 4.3.2). This procedure relies on the assumption that the search space is similar for all faces and environmental conditions. Cootes et al. show that the a-priori predicted gradient is only a good approximation of the real gradient for consistent training and application scenarios and within a limited range from the optimum [27].

This assumption was introduced in order to achieve a remarkable reduction of the computational effort. An online optimization scheme would require a comparably dense sampling of the actual multi-dimensional search space to find a deep local or ideally global optimum. Although this online coefficient optimization is a computationally intensive process, the quality of the convergence can be drastically increased by using the "real" gradient. Additionally, the advances in CPU and GPU developments within the last decade, equip modern consumer hardware with sufficient computational power to perform this task in reasonable time scales - assuming a correspondingly optimal software implementation.

This section will cover several approaches to an online AAM coefficient optimization. For a proper introduction in the principles and theory of non-linear optimization strategies please refer to [56].

Each of the online coefficient optimization algorithms tries to minimize E($\boldsymbol{h}$) =

5.3.1 Gradient Descent

$\frac{1}{2}||r(h)||^2$ (see eq. 4.27) in respect to the AAM coefficient vector h. In this work we implemented and evaluated three different basic approaches, namely Gradient Descent, Grid Sampling, and the Nelder-Mead algorithm [83], which are presented in the following sections. Note that according to our experiments [32] it is strongly suggested to normalize each search space dimension i to its specific variance, i.e. $\lambda_{c,i}$ for PCA-AAMs and $\sigma_{c,i}^2$ (eq. 5.61 for NMF-AAMs). Since the range of reasonable values for the AAM coefficients was found to be within a deviation of three times the standard deviation from the global mean (see section 4.2.3), the search space has a co-domain of ± 3 in each dimension. This is helpful to get a relation for the values of step width α and the sampling densities.

5.3.1 Gradient Descent

The update rule in a Gradient Descent optimization approach for $h_{[n]}$ at iteration n can in general be written as

$$h_{[n+1]} = h_{[n]} + \alpha_{[n]} \delta h_{[n]} \tag{5.64}$$

The algorithm starts with an initial estimation $h_{[0]}$ and a step width α. Hereby $\delta h_{[n]}$ constitutes a multi-variate gradient consisting in a direction and a steepness. The optimization task strives to find the global minimum with as few iterations as possible. Furthermore it should be able to avoid getting stuck in local minima. Due to the high dimensionality of the optimization problem ($h \in \mathbb{R}_c^\mu$ with $\mu_c \approx 30$) and the high computational costs of performing one iteration (esp. texture synthesis and warping), we implemented the estimation of the gradient $\delta h_{[n]}$ at the position $h_{[n]}$ as follows.

In accordance to the offline optimization described in section 4.3.4 it is suggested to update $\alpha_{[n]}$ between two consecutive iterations with one of the following decay terms. One example multiplies α with a fixed factor $\beta < 1$ after each iteration so that

$$\alpha_{[n]} = \alpha \cdot \beta^n \tag{5.65}$$

Another approach to update the step size is setting

$$\alpha_{[n]} = \frac{\alpha}{(1 + n \cdot \gamma)} \text{ with } \gamma > 0 \tag{5.66}$$

Our experiments have shown that the AAM shows generally better convergence with the decay term of equation 5.65 and an initial step width $\alpha_{[0]} \approx 1.5$ and $\beta \approx 0.65$.

5.3.1.1 Independence of Dimensions

Firstly, we assume an independence of each dimension in the search space, apart from the objective function of course. Thus, each AAM model coefficient is optimized independently:

$$h_{[n+1],(i+1)} = h_{[n],(i)} + \alpha_{[n]} \delta h_{[n],(i+1)} \tag{5.67}$$

According to this equation, the gradient $\delta h_{[n],(i+1)}$ is estimated only for the dimension $(i+1)$. The coefficient vector $h_{[n+1],(i+1)}$ of the next iteration $[n+1]$ is instantly updated with $\delta h_{[n],(i+1)}$. In this approach the vectors $h_{[n+1],0}$, $h_{[n+1],1}$, $\cdots$, $h_{[n+1],i}$, $\cdots$, $h_{[n+1],(\mu_c-1)}$ are computed successively. Hence, the estimation of the gradient in the dimension $(i+1)$ profits by the incorporated knowledge about the previous dimensions. This is especially superior to a single update by a multi-variate gradient per iteration, since the first dimensions with the highest Eigenvalues $\lambda_{c,0} \geqslant \lambda_{c,1} \geqslant \cdots \geqslant \lambda_{c,i} \geqslant \cdots \geqslant \lambda_{c,(\mu_c-1)}$ cover high variances and entail correspondingly significant changes in the search space of an PCA-AAM.

5.3.1.2 Interval Sampling

For the gradient estimation as well as for the Grid Sampling (see section 5.3.2) there are several methods of sampling the surrounding of the current position $h_{[n]}$ in the dimension i with P sample points.

In the first variant, i.e. an equal distribution of the sample points, their positions $h_{[n],i}^{(p)}$ are obtained from

$$h_{[n],i}^{(p)} = h_{[n],(i-1)} + \tau(p) \tag{5.68}$$

where

$$\tau(p) = p \cdot \frac{\gamma_{[n]}}{P} \qquad p \in \mathbb{N} \cap \left[-\frac{P}{2}; \frac{P}{2}\right] \setminus \{0\} \tag{5.69}$$

An exponential distribution of the sample points can be implemented by

$$\tau(p) = sign(p) \left[\left(\gamma_{[n]}+1\right)^{\frac{2|p|}{P}} - 1\right] \tag{5.70}$$

Unlike the equal distribution, this allows on the one hand a more dense search space sampling in the direct surrounding of the current position and on the other hand a more loose sampling of remote areas with the same amount of sample points. However, this behavior is solely advantageous at high numbers of sample points, e.g. $P > 8$, and large γ values, e.g. $\gamma > 1.5$.

As third possibility we implemented and tested a random distribution of P sample points within an interval of $\left[-\gamma_{[n]}/2; \gamma_{[n]}/2\right]$ which finally turned out to produce the best results.

In accordance with the step width $\alpha_{[n]}$, the interval size $\gamma_{[n]}$ should also be updated with a decay term. Here, we implemented the same approaches as for $\alpha_{[n]}$ (see equations 5.65 and 5.66).

5.3.1.3 Gradient Estimation

For each of the P sample points $h_{[n],i}^{(p)}$ around $h_{[n],(i-1)}$ a difference quotient in dimension i can be computed as approximation of the gradient by

$$\nabla \mathrm{E}_i^{(p)} \approx \frac{\mathrm{E}\left(h_{[n],(i-1)} + \tau(p)\right) - \mathrm{E}\left(h_{[n],(i-1)}\right)}{\tau(p)} \tag{5.71}$$

5.3.2 Grid Sampling

Instead of these one-side differences, it can be numerically more stable to rely on central differences. This would require twice as much warping and synthesis operations per coefficient. However, here the symmetry of the sample point distribution of equations 5.68 and 5.69 can be exploited. The central difference quotient provides an estimation of the gradient by

$$\nabla \mathrm{E}_i^{(p)} \approx \frac{\mathrm{E}\left(\boldsymbol{h}_{[n],(i-1)} - \tau(p)\right) - \mathrm{E}\left(\boldsymbol{h}_{[n],(i-1)} + \tau(p)\right)}{2\tau(p)} \quad (5.72)$$

Based on these two variants we obtain a difference quotient for each of the P sample points. Hence, we investigated several variants for the estimation of the gradient from the difference quotients, i.e. an averaging over the entire interval, an independent averaging on both sides of the sampling interval, and a maximum search. The averaging is usually conducted in order to smooth high-frequency noise in the search space. Our experiments [32] discovered that the error function $\mathrm{E}(\boldsymbol{h})$ and hence the search space of the AAM coefficient optimization shows just a negligible jitter. With this knowledge it is not surprising that the simplest approach, i.e. maximum search, lead to the best results.

The termination condition for the Gradient Descent optimization bases on the change of the minimal error function value

$$\Delta \mathrm{E}\left(\boldsymbol{h}_{[n+1]}\right) = \mathrm{E}\left(\boldsymbol{h}_{[n]}\right) - \mathrm{E}\left(\boldsymbol{h}_{[n+1]}\right) \quad (5.73)$$

If

$$0 \leqslant \Delta \mathrm{E}\left(\boldsymbol{h}_{[n+1]}\right) < \epsilon \quad (5.74)$$

the algorithm is considered to be converged and terminates.

5.3.2 Grid Sampling

From the implementation view, the Gradient Descent and Grid Sampling approaches are very similar. The major difference consists in the neglect of the multiplication of the delta vector $\delta \boldsymbol{h}_{[n]}$ with the step width $\alpha_{[n]}$ in equation 5.64. Thus, this equation reads

$$\boldsymbol{h}_{[n+1]} = \boldsymbol{h}_{[n]} + \delta \boldsymbol{h}_{[n]} \quad (5.75)$$

Still, $\gamma_{[n]}$ is relevant for the determination of the sampling interval and the sampling points as described in section 5.3.1.2. Grid Sampling provides for selection of the best of the P sample points $\boldsymbol{h}_{[n],i}^{(p)}$ which leads to the minimal error energy. Hence, unlike Gradient Descent, the updated coefficient vector $\boldsymbol{h}_{[n],i}$ lies directly on one of the sampled search space positions. Therefore, the values for P, $\gamma_{[0]}$, and β are in a different codomain for the Grid Sampling algorithm. This approach requires a higher number of sample points, wider sampling intervals, and a softer decay of the interval size.

The Grid Sampling incorporates the same termination condition as the Gradient Descent as described by the equations 5.74 and 5.73.

Again in contrast to the Gradient Descent algorithm, the exponential distribution of the sample points in the search interval is superior to an equal or random distribution, whereas a considerable number of points is required.

5.3.3 Nelder-Mead or Simplex Optimization

At both approaches mentioned above, Gradient Descent and Grid Sampling, the error function $\mathrm{E}(\boldsymbol{h})$ has to be evaluated $P \cdot \mu_c$ times per iteration n with typical values of $P \approx 10$ and $\mu_c \approx 30$. This corresponds to a number of $P \cdot \mu_c$ calculations of the difference image $\boldsymbol{r}(\boldsymbol{h}) = A \circ W(\boldsymbol{p}, \boldsymbol{s}(\boldsymbol{h}_c)) - \boldsymbol{t}(\boldsymbol{h}_c)$ which requires the computationally expensive tasks of texture warping from the original image and texture synthesis (see section 4.3.1). However, finally just the information of μ_c sample points, i.e. $\sim 10\%$ of the gained knowledge about the search space, is actually saved for the next iteration.

This is the decisive advantage of the Simplex Optimization introduced by J.A. Nelder and R. Mead in [83] as it is described in the following. Let the optimization search space be defined in $\mathbb{R}^{d+1}$. In case of AAMs the dimensionality adds up to $d = \mu_c + 1$, i.e. μ_c coefficients in $\boldsymbol{h}$ plus the one dimensional objective function $\mathrm{E}(\boldsymbol{h})$. Although in our case $d = \mu_c$ holds, we stick to the notation with d for an improved readability.

The central element of this method constitutes a d-*Simplex* which is a $(d+1)$ dimensional analogue of a triangle. In general, a simplex is the convex hull of a set of $(d+1)$ affinely independent points in some Euclidean space of dimension d or higher. Illustratively, *affinely independent points* are a set of points such that no m-plane contains more than $(m+1)$ of them. Such points are said to be in general position.

The initial d-simplex is positioned with the initial estimation $\boldsymbol{h}_{[0]}$ as its center of gravity and vertices of a distance of 1 from the center. In our case, the $(d+1)$ vertices are described by the coefficient vectors $\boldsymbol{h}^{(p)}$. According to our experiments the construction of the initial simplex has just marginal influence on the convergence properties as long as it is not chosen too small. Since the simplex is predominantly contracted during the optimization run, the initial simplex should not fill less than one fifth of the search space dimensions.

Now, at each vertex of the simplex the objective function is evaluated. Hence they are ordered by the value of their objection function in ascending order:

$$\mathrm{E}\left(\boldsymbol{h}^{(0)}\right) \leqslant \mathrm{E}\left(\boldsymbol{h}^{(1)}\right) \leqslant \cdots \leqslant \mathrm{E}\left(\boldsymbol{h}^{(d)}\right) \tag{5.76}$$

Let $\overline{\boldsymbol{h}}$ be the center of gravity of all points except the worst $\boldsymbol{h}^{(d)}$. Subsequently, the point $\boldsymbol{h}^{(d)}$ is reflected at $\overline{\boldsymbol{h}}$ according to

$$\boldsymbol{h}_r = \overline{\boldsymbol{h}} - \rho_r \left(\boldsymbol{h}^{(d)} - \overline{\boldsymbol{h}}\right) \tag{5.77}$$

We compute the error function at the new position $\mathrm{E}(\boldsymbol{h}_r)$.

This stage requires a case differentiation:

5.3.3 Nelder-Mead or Simplex Optimization

- $E(h_r) < E(h^{(0)})$, i.e. the new point is "better" than all vertices of the simplex. In this case an *expansion* is performed (see equation 5.78).

- $E(h^{(p)}) \leqslant E(h_r) < E(h^{(p+1)})$ with $p > 1 \vee p < d$, i.e. the overall position of the simplex was improved but no new "best" sample point could be found. In this case a *contraction* is performed (see equation 5.80).

- If the new point h_r is worse than all vertices before, a *shrink* is performed (see equation 5.81).

Since in the first case the reflection of the worst vertex leads to a new minimum, it is assumed that the direction of the reflection could point to the searched minimum. Thus, the new simplex is expanded in this way to the point h_e by

$$h_e = h_r + \rho_e \left(h_r - \overline{h}\right) \tag{5.78}$$

If $E(h_e) < E(h_r)$ the new simplex is described with the vertices

$$\left(h_e, h^{(0)}, h^{(1)}, ..., h^{(d-1)}\right) \tag{5.79}$$

Otherwise, h_r would be applied instead of h_e. A re-ordering of the vertices in accordance with equation 5.76 prepares the vertex for the next iteration.

In case of the new point h_r queues within the other vertices with respect to its error function, this indicates that the performed reflection does not point in the direction of the minimum. Therefore, the simplex is contracted by

$$h_c = \overline{h} + \rho_c \left(\overline{h} - h^{(d)}\right) \tag{5.80}$$

Again, if $E(h_c) < E(h_r)$ the new simplex is described with the vertices $\left(h_c, h^{(0)}, h^{(1)}, ..., h^{(d-1)}\right)$, otherwise applying h_r. The iteration is finalized with the re-ordering following equation 5.76.

When the result of the reflection shows an even higher error function $E(h_r)$, it can be assumed that the currently best vertex $h^{(0)}$ of the simplex lies close to or at least in the direction of the minimum. Hence the entire simplex is shrunken around this point and all other d points are re-computed:

$$h_s^{(i)} = h^{(0)} + \rho_s \left(h^{(i)} - h^{(0)}\right) \tag{5.81}$$

Unfortunately, after a shrink step the objective function must be evaluated for all moved vertices. Consequently this is the most computationally expensive possibility. Luckily, it is required very rarely according our experience during AAM coefficient optimization. It can appear especially at the end of the process, when the simplex converges around the minimum.

Eventually, the ordering of the vertices with respect to the objective function is performed again. Each Nelder-Mead iteration ends with the validation of the termination condition:

$$\Delta \overline{h^{(d)}} = \left\|\overline{h_{[n-1]}^{(d)}} - \overline{h_{[n]}^{(d)}}\right\|^2 < \epsilon \tag{5.82}$$

Here, $\overline{\boldsymbol{h}}^{(d)}$ denotes the center of gravity of the entire simplex in contrast to $\overline{\boldsymbol{h}}$ where only the best $(d-1)$ vertices are involved. During the Nelder-Mead optimization, the simplex tends to collapse around the found minimum. If finally the displacement of the center between two consecutive iterations falls under a certain threshold ϵ, the simplex got so small that a performed contraction or shrink had just marginal influence on the centroid of the simplex and thus no relevant impact on the found minimum. Since the displacement fell under the threshold, it is assumed that accordingly the error energy does not face further improvement and the optimization algorithm is considered to be converged.

The co-domain of all relevant factors ρ is $]0;1[$. These factors are valuable for the adjustment of the simplex behavior in the search space. For all ρ, a value close to 1 leads to a rough sampling and an agile simplex while a value close to 0 performs a denser sampling of the search space and keeps the simplex starchy. Please note that the factors ρ_e and ρ_c are directly dependent on ρ_r. Typical values are $\rho_r = 1$, $\rho_e = \rho_r = 1$, $\rho_c = 0.5\rho_r = 0.5$, and $\rho_s = 0.5$.

Eventually, the Nelder-Mead or Simplex Optimization algorithm is capable to preserve and incorporate the knowledge about the search space gained from the sample point from iteration $[n]$ to iteration $[n+1]$, except for the rare shrink case. Since each sampling point is highly valuable not just with respect to its information but especially with respect to its computational effort, the Nelder-Mead is highly superior in comparison to all other previously introduced approaches. However, its convergence properties around the minimum thwart this algorithm. An increase of the Δ-threshold ϵ as countermeasure however instantly delivers inferior optimization quality than a Grid Sampling with random sample point distribution.

A comprehensive discussion of the Nelder-Mead optimization strategy is given by K.E. Vurgin in [116]. He covers the convergence properties, and the well-known combination with the Simulated Annealing approach for search spaces with distinctive local minima.

5.3.4 Conclusion

The three introduced optimization strategies primarily differ in the computational complexity of an optimization run. The Gradient Descent as well as the Grid Sampling usually require 5 to 10 sample points for each dimension whereas the evaluation of each sample point needs an AAM synthesis. Consequently, the update of the complete coefficient vector with about 30 dimensions invokes up to 300 AAM syntheses. In contrast to these approaches, the Simplex algorithm updates the entire coefficient vector in each iteration and is therefore approximately 10 times faster than the other strategies. The only expensive operation is the shrink step of the Simplex. This appears when the Simplex melts into the found minimum and reduces the computational advantage against the Grid Sampling. The evaluations in chapter 6 show clearly that the Nelder-Mead or Simplex algorithm constitutes the best trade-off between computational effort and re-synthesis quality. Throughout, this optimization strategy is significantly superior to the standard optimization based on a predicted gradient.

5.4 GPU-Accelerated Active Appearance Models

The previous section has already given examples for the runtime complexity of several AAM coefficient optimization algorithms. We have identified the objective function from eq. 4.27 as most critical section of the algorithm. Basically there are two major aspects worthwhile to be discussed: the warping transformation W and the AM texture synthesis. Besides these two expensive operations the texture alignment A and difference energy calculation also have a serious impact on the overall runtime of the AAM coefficient optimization algorithm. Our goal is to optimize these aspects of AAMs by accelerating our implementation via GPU-based routines.

5.4.1 Warping

This section discusses the warping transformation W as introduced in section 4.1.2. We incorporate GPU-accelerated rendering operations based on OpenGL 2.0 [102] and multiple extensions to the standard. This section covers the theory and concepts behind our software module "WarpingEngine2" [84] which provides a broad interface and allows a comfortable integration in arbitrary C/C++ architectures.

5.4.1.1 Geometry

The goal of a warping operation is to geometrically deform an arbitrary region of an image and morph it from one set of scattered 2D vertices $\mathcal{X}$ to another set of scattered 2D vertices $\mathcal{X}'$. These vertices are represented as 3D homogeneous coordinates. A shape deformation on a texture in this context requires several items:

- A source image p that is downloaded to GPU texture memory.

- $|\mathcal{X}|$ source vertices $\mathcal{X} = \{x_i = (x_{0_i}, x_{1_i}, 0, 1)^T \mid 0 \leqslant i < |\mathcal{X}|\}$, defining the 2D geometry of an arbitrary region on the source image p, that shall be geometrically deformed.

- The same number of $|\mathcal{X}|$ destination vertices $\mathcal{X}' = \{x'_i = (x'_{0_i}, x'_{1_i}, 0, 1)^T \mid 0 \leqslant i < |\mathcal{X}|\}$ which define the shape the image region shall be warped into.

- A triangle index list $\ominus$ that describes the geometry itself. One triangle $\psi_i \in \ominus$, $0 \leqslant i < |\ominus|$ is defined as 3-element vector $\psi_i = (\psi_{0_i}, \psi_{1_i}, \psi_{2_i})^T$ with $\psi_{0_i}, \psi_{1_i}, \psi_{2_i} \in \{0, \ldots, |\mathcal{X}| - 1\}$ and $\psi_{0_i} \neq \psi_{1_i} \neq \psi_{2_i}$ correspond to the three indices of the vertices which describes the triangle ψ_i. Triangles are generated using a modified DeWall triangulation algorithm [32]. This algorithm generates a planar set of triangles $\ominus$ for a given set of scattered vertices in 2D space each triangle holding the Delaunay criterion [29].

The geometry of a shape rendered by an OpenGL pipeline is made up of the source vertices $\boldsymbol{x}_i \in \mathcal{X}$, the vertex color $\boldsymbol{\chi}_i = (\chi_{0_i}, \chi_{1_i}, \chi_{2_i}, \chi_{3_i})^T \in \chi$, the normals $\boldsymbol{n}_i = (n_{0_0}, n_{1_0}, n_{2_0})^T \in \mathcal{N}$, and the destination vertices $\boldsymbol{x}'_i \in \mathcal{X}'$. All of these values are stored interleaved in the OpenGL vertex array $\boldsymbol{b}$ which is defined as

$$\boldsymbol{b} = (\underbrace{x_{0_0}, x_{1_0}, 0, 1}_{x_0}, \underbrace{\chi_{0_0}, \chi_{1_0}, \chi_{2_0}, \chi_{3_0}}_{\chi_0}, \underbrace{0, 0, 0}_{n_0}, \underbrace{x'_{0_0}, x'_{1_0}, 0, 1}_{x'_0}, \ldots, \tag{5.83}$$

$$\underbrace{x_{0_{|\mathcal{X}|-1}}, x_{1_{|\mathcal{X}|-1}}, 0, 1}_{x_{|\mathcal{X}|-1}}, \underbrace{\chi_{0_{|\mathcal{X}|-1}}, \chi_{1_{|\mathcal{X}|-1}}, \chi_{2_{|\mathcal{X}|-1}}, \chi_{3_{|\mathcal{X}|-1}}}_{\chi_{|\mathcal{X}|-1}}, \underbrace{0, 0, 0}_{n_{|\mathcal{X}|-1}}, \underbrace{x'_{0_{|\mathcal{X}|-1}}, x'_{1_{|\mathcal{X}|-1}}, 0, 1)^T}_{x'_{|\mathcal{X}|-1}}$$

The vertex normals $\boldsymbol{n}_i \in \mathcal{N}$ are unused and set to zero as they are only important for lighting calculation which is not required for straight warping. The order of the elements in the vertex buffer are optimized for the OpenGL rendering pipeline and accessed via a vertex buffer object [67, 49] as interleaved vertex array [103]. Figure 5.12 illustrates the rendering pipeline of the WarpingEngine2. An arbitrary number of k shapes can be combined to produce one warping result in the framebuffer.

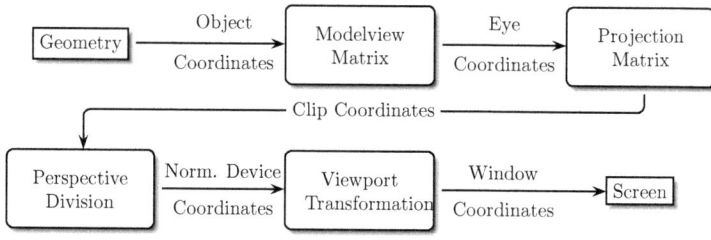

Figure 5.12: WarpingEngine2 rendering pipeline

The source vertices $\mathcal{X}$ directly correspond to the AAM landmarks $\boldsymbol{s}_i$ defined on the input image $\boldsymbol{p}_i$ and are loaded as texture coordinates into the OpenGL pipeline.

The destination vertices are loaded as 2D-geometry into the OpenGL pipeline and define the object that is going to be rendered. Parts of the source image $\boldsymbol{p}$ (defined by the source vertices) are mapped as texture onto the geometry defined by the destination vertices.

5.4.1 Warping

The bounding box [1] of the destination vertices is later used to adjust the projection so it perfectly fits the destination vertices (see figure 5.13).

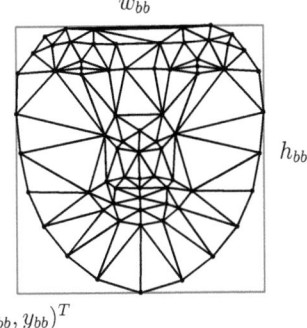

Figure 5.13: The bounding box of an AAM shape s (destination vertices)

5.4.1.2 Projection

As 2D warping is desired only, it is reasonable to set an orthographic projection, eliminating perspective artifacts and keeping parallel lines parallel after projection. It is not distinguished between objects which are near to the camera or far away from the observer – except that their rendering order is determined by their z-coordinate.

Identity For a graphical illustration of the scene setup see fig. 5.14. The observer is viewing the scene into the positive z-axis direction. The origin is located in the lower left corner of the framebuffer. The viewing frustrum is visualized as a gray box where areas outside are clipped away. The near clipping plane is located at $z = \nu$ and the far clipping plane at $z = \varphi$.

The *viewport* is adapted to fit the extends of the framebuffer and the projection is modified so the observer (camera) perfectly covers the bounding box of the destination vertices $\mathcal{X}'$. Assuming it has a width of w_{bb} and a height of h_{bb} units and its position is $(x_{bb}, y_{bb})^T$ within the world coordinate system. If the framebuffer has a resolution of $w_{fb} \times h_{fb}$ pixels, with ν and φ corresponding to the near and far clipping planes, the initial projection assuming that the left bottom origin is set to $(0,0)^T$ is defined as

$$\begin{bmatrix} \frac{2}{w_{fb}} & 0 & 0 & -1 \\ 0 & \frac{2}{h_{fb}} & 0 & -1 \\ 0 & 0 & -\frac{2}{\varphi-\nu} & -\frac{\varphi+\nu}{\varphi-\nu} \\ 0 & 0 & 0 & 1 \end{bmatrix} \quad (5.84)$$

---

[1] The bounding box of a scattered set of vertices is the minimal rectangular box that contains all of the vertices. The bounding box is perpendicular to the x and y axis of the coordinate system

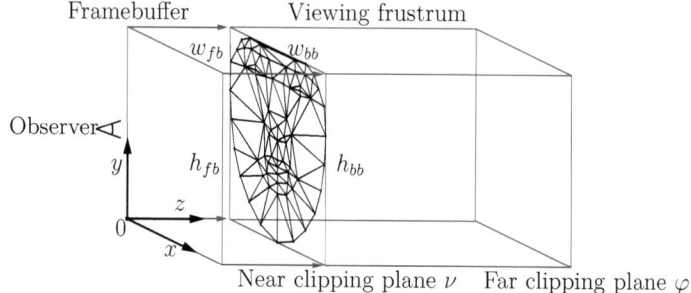

Figure 5.14: WarpingEngine2 scene setup: viewport, projection and viewing frustrum

This projection is scaled by $\frac{w_{fb}}{w_{bb}}$ in x-direction, $\frac{h_{fb}}{h_{bb}}$ in y-direction and respectively translated by $(-x_{bb}, -y_{bb})^T$. The final projection matrix $\boldsymbol{P}$ can be calculated by applying these transformations to the matrix defined in eq. 5.84, producing

$$\boldsymbol{P}_I = \begin{bmatrix} \frac{2}{w_{bb}} & 0 & 0 & -\frac{w_{bb}^2 + w_{fb}}{w_{bb}} \\ 0 & \frac{2}{h_{bb}} & 0 & -\frac{h_{bb}^2 + h_{fb}}{h_{bb}} \\ 0 & 0 & -\frac{2}{\varphi - \nu} & -\frac{\varphi + \nu}{\varphi - \nu} \\ 0 & 0 & 0 & 1 \end{bmatrix} \quad (5.85)$$

The projection transformation $\boldsymbol{P}_I$ is applied during rendering operations to make sure that the destination vertices optimally fill out the whole framebuffer.

Translation, Rotation, and Scaling It is equivalent to either transform the projection or instead transform the geometry applying the inverse transformation. The projection transformation is only used to adapt the camera to the bounding box of the geometry. Section 5.4.1.5 explains how the location, scaling and rotation of the geometry itself is modified.

5.4.1.3 Source image

The source image $\boldsymbol{p}$ is stored in the GPU texture memory as 2D texture. The OpenGL 2.0 standard normally requires textures to have power-of-two dimensions. This limitation is eliminated by using the ARB_texture_non_power_of_two [98] extension. This extension has several advantages over the ARB_texture_rectangle or NV_texture_rectangle extensions which implement a very similar concept [61]: Usage is natively supported, no special texture type is introduced and all advanced texture features such as filtering or texture borders are available.

As texture allocation is an expensive operation, texture memory is recycled where applicable. The texture layout and usage scenario is shown in fig. 5.15. To re-use texture memory, a rectangular virtual region is mapped onto a physical area in texture memory. Initially a texture is always generated exactly matching the

5.4.1 Warping

requested resolution. Successive iterations may re-use the allocated memory area if it is large enough to hold the texture data, allowing to use only a subregion of it. If the physical texture is not large enough to hold the new texture, it is enlarged. Thus we must always assume, that virtual and physical texture dimensions differ, with the virtual texture resolution being equal or less than the physical.

5.4.1.4 Source vertices

The source vertices $x_i \in \mathcal{X}$ are mapped on normalized texture coordinates $\hat{x}_i \in \hat{\mathcal{X}}$. One has to distinguish between *physical* and *virtual* texture coordinates where both have their origin in the upper left corner of the texture. Physical texture coordinates are always normalized in respect to the physical texture size of $w_t \times w_h$ where w_t and w_h are measured in pixels. Furthermore a region of interest (*abbrev.* 'ROI') can be defined on the image p that has a dimensionality of $w_r \times h_r$ pixels and its origin at $(x_r, y_r)^T$ pixels. A region of interest is essential for all AAM optimization algorithms for pose adjustments being calculated relative to the size of the head. This makes the whole AAM coefficient optimization algorithm resolution independent. For instance, a translation of -0.5 in x-direction moves the texture coordinates (the source vertices) half the width of the head.

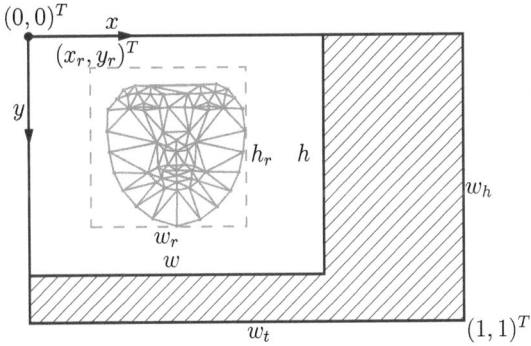

Figure 5.15: WarpingEngine2 texture setup: virtual and physical texture

As transformations shall be applied relative to the source image ROI, a mapping between the ROI and the physical texture coordinates has to be found, completely hiding the virtual texture paradigm from the programmer.

Identity The source vertices $\mathcal{X}$ given in pixel coordinates relative to the upper left corner $(x_r, y_r)^T$ of the source image ROI have to be normalized. Setting the texture identity transformation to

$$T_I = \begin{bmatrix} \frac{1}{w_t} & 0 & 0 & x_r \\ 0 & \frac{1}{w_h} & 0 & y_r \\ 0 & 0 & 1 & 0 \\ 0 & 0 & 0 & 1 \end{bmatrix} \quad (5.86)$$

transforms the source vertices $x_i = (x_{0_i}, x_{1_i}, 0, 1)^T \in \mathcal{X}$ into the normalized texture coordinates $\hat{x}_i \in \hat{\mathcal{X}}$ with

$$\hat{x}_i = T x_i \quad (5.87)$$
$$= \left(\frac{x_{0_i}}{w_t} + x_r, \frac{x_{1_i}}{w_h} + y_r, 0, 1 \right)^T \quad (5.88)$$

Translation To translate the normalized texture coordinates $\hat{x}_i \in \hat{\mathcal{X}}$ relative to the size of the source image ROI by $(t_x, t_y, t_z)^T$, the following transformation must be applied:

$$\hat{x}_i' = S^{-1} L S \hat{x}_i \quad (5.89)$$

with L being a default translation matrix and S a scaling matrix that scales by $(w_r, h_r, 1)^T$. The final translation matrix can then be written as

$$T_t = \begin{bmatrix} 1 & 0 & 0 & \frac{t_x}{w_r} \\ 0 & 1 & 0 & \frac{t_y}{h_r} \\ 0 & 0 & 1 & t_z \\ 0 & 0 & 0 & 1 \end{bmatrix} \quad (5.90)$$

Rotation, Scaling To rotate or scale the source vertices $\mathcal{X}$ on the source image p default homogeneous coordinate transformations are applied. Prerequisites such as translation of the texture coordinate space origin into the center of rotation must be met accordingly.

5.4.1.5 Destination vertices

The destination vertices $\mathcal{X}'$ are passed into the OpenGL pipeline as specified. The projection is adapted to fit their bounding box (see fig. 5.13), so no modification of the model identity transformation $M_I = I$ is necessary.

To achieve resolution and model independent translation relative to the size of the bounding box of the destination vertices by $(t_x, t_y, t_z)^T$, the scaled model translation matrix M_t is defined as

$$M_t = \begin{bmatrix} 1 & 0 & 0 & \frac{t_x w_{bb}}{w_{fb}^2} \\ 0 & 1 & 0 & \frac{t_x h_{bb}}{h_{fb}^2} \\ 0 & 0 & 1 & t_z \\ 0 & 0 & 0 & 1 \end{bmatrix} \quad (5.91)$$

Rotation and scaling is applied unmodified by either a general rotation matrix R or a scaling matrix S.

5.4.1.6 Warping

The warping process takes the destination vertices $\mathcal{X}'$ as geometry and maps the source image as texture onto it while it is being rendered. The source vertices $\mathcal{X}$ define the texture coordinates and thus control how the texture is going to be mapped onto the rendered geometry.

During rendering several blending attributes may influence how the final fragments are combined. Let $\chi_t \in \mathbb{R}^4$ be the texel color, $\chi_f \in \mathbb{R}^4$ the fragment color, and $\chi_v \in \mathbb{R}^4$ the interpolated vertex color. The operator $\odot$ is the element-wise vector multiplication. In this case each of the vector elements is simply multiplied with the scalar s. Generally a color is described by a 4-element vector, containing one element red, green, blue, and an alpha transparency value, respectively. Then the final fragments color is calculated by $\chi_f = \chi_t \odot \chi_v$ which modulates the texture color by the vertex color χ_v of the shape. This allows for blending the rendered shape with other shapes beneath or with the background.

Currently triangles are not sorted back to front so it might occur that they overlap and produce artifact as shown in figure 5.16. In the future this could be solved by either manually or automatically adding depth information to the destination vertices. However, as the vertices of a human face in an Appearance Model usually do not produce such artifacts or only under extreme conditions, it can be ignored.

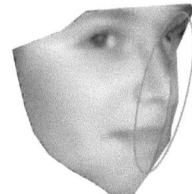

Figure 5.16: Incorrectly warped texture with overlapping triangles on the right edge of the face

5.4.2 Coefficient Optimization

This chapter describes fundamental concepts and algorithmic design issues of GPU-AAMs. Implementation specific details, data structures, and how GPU-AAMs are actually implemented can be found in [32].

AAM coefficient optimization necessitates many costly operations. It is primarly any texture related task that renders the overall method expensive. This is why we emphasize out efforts on optimizing these costly operations.

5.4.2.1 Minimizing traffic

If only warping is performed on the GPU as in previous implementations [32], it is necessary to transfer the warped texture back into CPU memory for each iteration during coefficient optimization. Assuming an RGB color texture with a resolution of 128 × 128px these are about ∼ 192KB for each warped texture. If doing $n = 10$ Gauss-Newton iterations ∼ 1.875MB have to be transferred for optimizing one image frame. Besides the occurring traffic it is also necessary to synchronize with the OpenGL pipeline as all OpenGL rendering operations must be finished, before the resulting frame can be read back from device into host memory. This destroys any asynchronism. Hence it is important to minimize traffic between host and device memory.

Figure 5.17 illustrates the CPU-based coefficient optimization. In this context the term CPU-based is not really correct, as warping is already performed on the GPU but as most of the optimization algorithm (texture synthesis, difference image calculation, error energy calculation) is performed by the CPU, we call this approach CPU-based AAM coefficient optimization. This demands a permanent read-back of the warped texture from device to host memory, indicated by the double arrow.

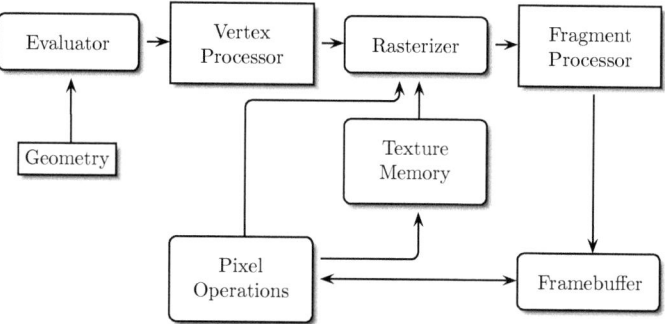

Figure 5.17: Schematic illustration of CPU-based coefficient optimization

In contrast to the CPU-based optimization our GPU-based approach is illustrated by figure 5.18. Most of the texture related, expensive operations are performed by the device, like warping, texture synthesis and difference image calculation. Only the shape synthesis is performed by the host.

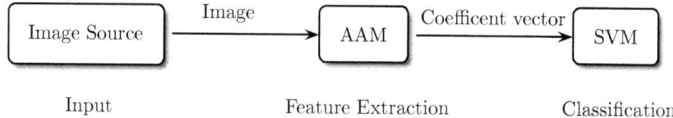

Figure 5.18: Schematic illustration of GPU-based coefficient optimization

The image that shall be optimized is downloaded only once to device memory and remains there during all iterations. The texture synthesis as well as the difference image calculation is performed by the device, eliminating the necessity

5.4.2 Coefficient Optimization

for reading back the warped result into host memory. The texture difference and energy calculation is also performed by the device and only one float value – the texture difference energy – is read back to host memory.

Especially sampling algorithms that calculate their gradient online highly profit from this strategy, as they require at least one sequence of warping-synthesis-difference per AAM coefficient per iteration which sums up to a huge number of warping, synthesis and texture difference energy calculations. For instance, a gradient descent with $n = 10$ iterations, a grid sampling with $P = 10$ samples for about $|h| \approx 30$ coefficients, requires 3000 single warping, texture and difference energy calculations *per image*.

5.4.2.2 Texture synthesis

The texture synthesis as introduced in eq. 4.26 is very expensive, as the matrix Q_t is huge. Assuming that a RGB color model with a texture resolution of 128×128px is used, the texture basis Q_t is defined as $Q_t \in \mathbb{R}^{49152 \times \mu_c}$. The matrix-vector product and vector sum in $t(h_c) = \bar{t} + Q_t h_c$ can largely benefit from the parallelism made available by GPUs. Hence, it is promising to do texture synthesis completely on the GPU. The matrix Q_t has only to be downloaded once to GPU memory and only the coefficient vector h_c which is very small (usually $\mu_c \approx 30$) compared to a whole texture has to be transferred each iteration. The result of the texture synthesis can directly reside in device memory used as operand during texture difference calculation to gain $r(h)$ and finally calculate the energy E(h). The texture synthesis is implemented via the CUBLAS SGEMV operation [86, 87]. The result resides in CUDA address space.

5.4.2.3 Warping and texture alignment

The geometric deformation of the original image texture p into the shape given by the mean shape $\bar{s}$ is also done by the GPU. The implementation package WarpingEngine2 is used to perform any warping operations during AAM coefficient optimization. The result of warping is rendered into texture memory and is left there for further processing. It is used as second operand during texture difference calculation to gain $r(h)$.

The warped result is mapped into CUDA address space as by now it is only available in OpenGL texture memory. It is copied into a PBO and mapped into CUDA accessible memory. The copy operation is acceptable as it is only a device-device copy that largely benefits from ultra-fast device memory[2].

As the warped texture $t_w = W(p, s(h_c))$ lies in a different domain than the synthesized texture t, it is necessary to align the warped texture t_w to the mean texture $\bar{t}$, applying the texture alignment transformation A from eq. 4.31. As the warped texture is available in CUDA after mapping, the alignment is implemented in CUBLAS.

---

[2] The NVIDIA G8x chipset has a peak memory transfer rate of about ~ 80GB/sec.

5.4.2.4 Difference energy

The objective function $E(\boldsymbol{h}) = \frac{1}{2}||A \circ W(\boldsymbol{p}, \boldsymbol{s}(\boldsymbol{h}_c)) - \boldsymbol{t}(\boldsymbol{h}_c)||^2$ that shall be optimized depends on the difference image between warped and synthesized texture. On this difference image the energy is calculated to finally produce a single floating point value that measures the convergence quality and is the target value of optimization.

Both operands, the warped as well as the synthesized texture are available in CUDA address space. Thus the difference and error energy is calculated using CUDA/CUBLAS.

5.4.3 Conclusion

Due to the very large amount of data and the resulting computational complexity of the image processing operations in AAMs, we developed strategies to transfer the most consuming tasks to the graphics processor (*abbrev.* 'GPU'). We have identified the objective function from eq. 4.27 as most critical section of the algorithm. Basically there are two steps which deserve the best possible acceleration: the warping transformation W and the AM texture synthesis. Besides these two expensive operations the texture alignment A and difference energy calculation also have a serious impact on the overall runtime of the AAM coefficient optimization algorithm. In this section we have presented the required data structures and routines allowing for a efficient processing of the AM data by the GPU plus a minimization of the data exchange traffic between the GPU memory and the CPU RAM. Depending on the resolution of the Appearance Model, our GPU based implementation provides an accelerated face analysis by factor 4 to 7. While the benefit is less significant at the offline optimization, strategies such as the Simplex (see section 5.3.3) with their high number of required face syntheses are boosted by GPU involvement.

5.5 Evaluation Measures for the Quality of AAM Re-synthesis

When it comes to the application of Active Appearance Models for the parametrization of faces, two scenarios shall be considered herein.

- *Face re-synthesis*: The target is to find an optimal reconstruction of the face by the AAM coefficients emphasizing a correct placement of the landmarks. Commercial applications for instance use the original face image with the automatically positioned landmarks to generate a Talking Head [15]. The quality of the re-synthesis is determined by the displacement of the automatically acquired landmarks from their ideal positions. However a comparison would require a manual annotation of the analyzed image. Finally, the quality is measured subjectively from the degree of realness of the re-synthesis by human perception.

5.5.1 Dataset Annotation

- *Information extraction*: Here, the parametrization of the unknown face by AAM coefficients shall serve as features with high entropy for a subsequent recognition of relevant properties of the face. The quality of re-synthesis is directly correlated with the recognition performance as we will see later on. However, the recognition performance is influenced by too many additional parameters, such as the chosen classification algorithm, its parameter setting, the data corpus, etc., and does not pay for a reliable quality measure.

After all, the quality of the AAM re-synthesis is subject to the human perception. Thus, an automatically computable measure shall be identified which emulates the perception of quality. Furthermore it should be independent from the number of landmarks and the resolution of the model in order to preserve comparability over different AAM parameter settings. Based on the most suitable measure, the re-synthesis performance of our implemented AAM variants is evaluated in section 5.5.3.

At first, a test set of images was created. Based on this set four different quality measures were implemented and evaluated. These are: error energy, spatial error energy, and inner vector product.

5.5.1 Dataset Annotation

We performed a random selection of 200 images from three different image databases: AR Database [71], FG-Net Aging Database [1], NI-Face Database [2]. These databases are described in section 6.2. Let the set of 200 images be denoted by $\mathcal{P}_Q$. The re-synthesis of the images $p_{q,i} \in \mathcal{P}_Q$ was conducted using the standard AAM approach with predicted gradient optimization (see section 4.3.2). For each database a specific AAM is applied which does not contain any images of $\mathcal{P}_Q$. Thus all analyzed images are unknown to the model. The quality of the re-synthesis was annotated by four persons in the scale of 1 (very good) to 5 (very bad). The inter-annotator agreement measured by the Fleiss' Kappa coefficient [39]

$$\kappa = \frac{\overline{P} - \overline{P_e}}{1 - \overline{P_e}} \tag{5.92}$$

resulted to $\kappa = 0.7$. It indicates the inter-annotator agreement with respect to the agreement to be expected by chance relative to the frequency of the class occurrences.

The annotation results for table 5.1 were gained by cropping of the best and the worst rating and computing the rounded average value out of the remaining two annotators. Table 5.5.1 shows the distribution of annotation classes.

Class number	1	2	3	4	5
Occurrences	23	27	53	51	46

Table 5.1: Distribution of convergence quality annotation from 1 (very good) to 5 (very bad)

Hence, we regard the task of finding a suitable quality measure as classification problem. The functions described in the following sections serve as *features* for a C4.5 Decision Tree which strives to find optimal thresholds as decision functions for classification.

5.5.2 Quality Measures

In the following sub-sections, various measures are introduced which all describe different properties of the similarity between the synthesized texture and the shape normalized original image (see section 4.3). In the subsequent section, all measures are applied and selected as features in a classification task to resemble the human perception as discussed above in 5.5.1.

5.5.2.1 Error Energy

All strategies for the AAM coefficient optimization described in this work include termination criteria which base on the final change of the error energy falling beyond a given threshold.

This error energy is obtained from equation 4.27. Independence from the model resolution is ensured by a normalization to the number of pixels c (compare equation 4.9). Thus, the relative error energy results to

$$\mathrm{E}(\boldsymbol{h}) = \frac{1}{2 \cdot c} ||\boldsymbol{r}(\boldsymbol{h})||^2 \qquad (5.93)$$

However, measuring the quality of convergence by examining the pure difference texture is suboptimal, as it does not ensure that the AAM actually found the correct location of the face within the input image [46]. Especially, the difference texture reveals only indirectly the placement of the shape landmarks.

Furthermore it averages over the entire texture. Therefore, a re-synthesis which for instance widely matches the eye region but fails the mouth region may produce the same error energy as a re-synthesis where for example the x-translation places the face too far aside. Additionally, a difference image showing a texture deviation over the entire face will again lead to a similar value of the error energy.

Considering these properties of the error energy as evaluation measure the poor classification result shown in section 5.5.3 is not astonishing.

5.5.2.2 Spatial Error Energy Distribution

In order to overcome the drawbacks of the pure error energy function (eq. 5.93) we investigated different similarity measures which explain the spread, accumulated coverage, and intensity, of areas with high texture difference values. For this purpose the difference image is converted to a binary image applying a difference threshold λ by means of all pixels with a difference $> \lambda$ are labeled with 1, all others with 0.

5.5.2 Quality Measures

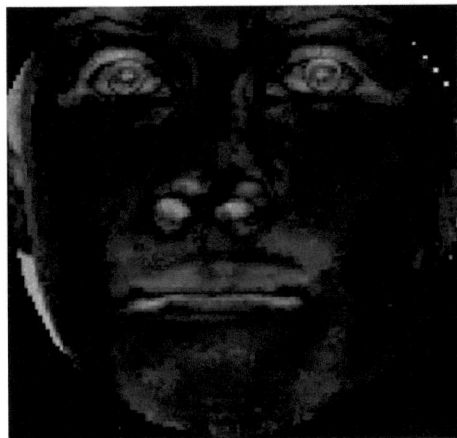

Figure 5.19: Example of an AAM difference image

A "blob detection" algorithm is applied on the binary difference image. This delivers a set of connected areas (blobs) of 1-pixels. Hence, for each blob b_i the x- and y-dimension $b_{x,i}$ and $b_{y,i}$ of the framing rectangle as well as the size in number of pixels $|b_i|$ is computed. In order to avoid noise effects, all blobs with a size $|b_i| < 1\% \cdot c$ are neglected for further considerations, where c is the number of pixels of the AAM texture. Let a number of I blobs be kept in this way. The average compactness of all blobs is determined by $\gamma = \frac{1}{I}\sum_i \frac{|b_i|}{b_{x,i} * b_{y,i}}$. The accumulated relative coverage of all blobs is obtained from $|B| = \sum_i |b_i|/c$. Further the averaged error energy over all blob-pixels and the relative sum of all blob pixels normalized by c is computed. Finally, the overall maximum pixel error value (i.e. brightest pixel in $r(h)$) is determined.

Eventually six additional parameters for the description of the difference image and thus of the re-synthesis quality have been extracted.

5.5.2.3 Inner Product

The previously presented measures exclusively operate on the difference image $r(h)$ between the synthesized texture and the shape normalized original image, as expressed in equation 4.30 and repeated here:

$$r(h) = A \circ W(p, s(h_c)) - t(h_c) \quad (5.94)$$

As known from geometry the length and orientation of two vectors can be compared by the *Inner Product* or *Dot Product* or *Scalar Product* as equivalents. We can consider the textures $t(h_c)$ and $A \circ W(p, s(h_c))$ as vectors of the same high-dimensional space but with possibly different orientations [110]. In order to obtain the angle Θ between the vector we must compute the Inner Product:

$$\Theta = \arccos \frac{[A \circ W(\boldsymbol{p}, \boldsymbol{s}(\boldsymbol{h}_c))] \cdot \boldsymbol{t}(\boldsymbol{h}_c)}{|A \circ W(\boldsymbol{p}, \boldsymbol{s}(\boldsymbol{h}_c))| \, |\boldsymbol{t}(\boldsymbol{h}_c)|} \qquad (5.95)$$

Unfortunately, due to the in average low error values and the high dimensionality of the space on the other hand, the angles Θ lie in a range too close to zero considering the computational accuracy. Thus, we neglect the arccos and instead set the quotient to the power of $\nu = 30$ to obtain a meaningful stretching in the domain between 0 and 1.

$$\Omega_T = \left(\frac{[A \circ W(\boldsymbol{p}, \boldsymbol{s}(\boldsymbol{h}_c))] \cdot \boldsymbol{t}(\boldsymbol{h}_c)}{|A \circ W(\boldsymbol{p}, \boldsymbol{s}(\boldsymbol{h}_c))| \, |\boldsymbol{t}(\boldsymbol{h}_c)|} \right)^{\nu} \qquad (5.96)$$

As the results presented in the subsequent section show, this measure resembles best the human perception of re-synthesis quality. Therefore, this measure is henceforth referred to as *Texture Similarity* with the notation Ω_T. The ideal re-synthesis would result in a value of $\Omega_T = 1$. The worse the similarity is, the more will the value decrease against 0.

The same procedure can be applied for measuring a *shape deformation* Ω_S. Here Ω_S is computed by the Inner Product of the mean shape $\overline{\boldsymbol{s}}$ (see section 4.1.1) and the shape synthesized during optimization by $\boldsymbol{s}(\boldsymbol{h}_c) = \overline{\boldsymbol{s}} + \boldsymbol{Q}_s \boldsymbol{h}_c$ (compare equation 4.25). It indicates the deviation of the synthesized shape from the mean shape. In practice this measure is less adequate to resolve very good and good re-synthesis quality but rather for detection of divergence of the AAM during coefficient optimization. Thus we applied this measure as additional termination or restart condition for the optimization strategy. It also allows for weeding out images with an exceedingly high shape deformation for a subsequent classification task, since such analysis will not provide an accurate parametrization of the analyzed object.

5.5.3 Evaluation of Quality Measures

The preceding sections introduced a considerable set of descriptors for the quality of AAM re-synthesis. All these measures were evaluated separately and as feature set. The target is to find measures and decision rules which are able to reproduce best the results of the human perception experiment in section 5.5.1. For this purpose a C4.5 decision tree [94] training and evaluation was performed in WEKA [121]. Applying the C4.5 on a single feature basically results in a set of n intervals for an n-class problem. The confusion matrices and the mean accuracy for the set of overall eight features are shown in table 5.2 (a). The classification accuracy of the error energy E($\boldsymbol{h}$) alone lead to poor 42%. The best single feature and just 2% below the set recognition performance turned out to be the Texture Similarity Ω_T which widely confirms the studies in [110].

Due to the low computational effort compared to the spatial error energy measures, the Texture Similarity was chosen to be the basis for all evaluations addressing the re-synthesis quality of the AAM analyses.

	1	2	3	4	5
1	7	9	4	3	0
2	3	20	4	0	0
3	1	3	39	8	2
4	1	0	1	5	4
5	0	2	2	3	29
Mean acc.:					**70%**

(a)

	1	2	3	4	5
1	5	12	2	3	0
2	2	18	0	5	2
3	0	5	41	12	0
4	0	1	2	42	6
5	0	3	0	13	30
Mean acc.:					**68%**

(b)

Table 5.2: Confusion matrices for C4.5 re-classification of all quality measures (a) and Texture Similarity only (b)

5.6 Summary

Apart from a detailed overview of the activities and research results in the area of Active Appearance Models, this chapter introduced an excerpt of the developed variants and improvements during our research. The contributions address all stages of the Active Appearance Model algorithm, namely the model generation, the coefficient optimization, and the evaluation of the re-synthesis quality. Additionally, the aspects of an efficient implementation based on GPU involvement are covered. When it comes to real life applications an implementation accelerated by graphics programming is mandatory to handle the immense computational effort. With the Non-negative Matrix Factorization a completely novel variant of Appearance Model generation is presented which can be subject of multiple further research activities. Standard AAMs suffer under the widely unrealistic precondition that the search space of AAMs is similar for all target objects under the whole range of variations in initializations, illuminations, and views. In order to overcome this constraints, we investigated several optimization strategies which operate without prediction schemes about the search space. Thereby, the Simplex strategy turned out to be powerful with respect to the computational efficiency and the quality of the re-synthesis. With this AAM extension the way to new applications and the fast acquisition of (semi-)automatically annotated training material is open.

Chapter 6

Application of Active Appearance Models to Face Analysis

All methodologies, theoretically introduced in the chapters 4 and 5 were implemented during numerous research works in conjunction with this work. The engineering and implementation of the FacE Analysis System FEASy (see section 1.2) attached great importance to the provision of an interface which allows for the application of all described AAM variants and the setting of virtually all corespondent algorithmic parameters. For instance, this XML-based interface provides the opportunity to perform an AAM generation at different resolutions, with PCA or NMF, for arbitrary image sets, and applying one of the various optimization strategies plus the setting of the specific parameters influencing the behavior of selected sub-methods. This demands for a multi-layered, object oriented, and highly flexible software architecture with a correspondingly high design and implementation effort.

After all, this is the basis for the manifold investigations of the peculiarities and capabilities of the Active Appearance Model algorithm and its derivatives. Still the evaluations presented in this chapter are a cutout of the possible issues to be examined around AAM performance. We try to clarify basic questions regarding, e.g. the coherencies of different parameters, the applicability to face analysis, and computational speed of the standard AAM, approach as well as the the novel abilities aroused by various parametric and algorithmic variations and extensions. This is conducted on the basis of several image databases for the generation of AAMs and the testing of the performance with respect to pattern recognition tasks on human faces. We consider person specific attributes such as the gender, identity, and age plus short-time variable properties like facial expression and head pose [63].

6.1 Classification Based on Results of the AAM Optimization

This section discusses the so far unaddressed issue of tangible utilization of the results obtained from an AAM analysis on an unknown image. Assuming that the AAM optimization provides a precise object re-synthesis and thus a specific parametrization by its model coefficients condensed in the coefficient vector h_c, it is still subject to research how the information, extracted and represented in this highly compact manner, shall be exploited to obtain relevant properties of the analyzed object.

Thereby, two possible approaches become apparent:

- Comparison of the re-synthesis quality of class specific AAMs:
  For each of the targeted classes a separate AAM is generated. The class assignment is performed by a maximum search over the four Texture Similarities produced by the four different AAM analyses.

- Statistical classification of the edited AAM coefficient vector:
  A statistical classifier is learned by edited coefficient vectors obtained from AAM analyses of a number of example images for the targeted classes. This trained classifier performs the class assignment on the edited coefficient vector of unknown images which have been analyzed by the AAM.

6.1.1 Classification based on class specific AAMs

As mentioned above this approach bases on the assumption that an AAM, generated from images of a certain class, will be capable to re-synthesize unknown images better than AAMs for other classes. To depict the procedure, imagine e.g. a facial expression recognition task with the four classes Smile, Frown, Yell, and Neutral. In this case four different AAMs must be generated for the four classes. Hence the analysis of one unknown image requires a processing applying those four AAMs. Finally the AAM showing the best re-synthesis with respect to a quality measure such as the Texture Similarity (see section 5.5.2.3) determines the class assigned to the image.

This approach faces multiple disadvantages regarding annotation and computational effort not to mention the classification performance: In order to allow for an adequate generalization, a considerable number of images is required for AAM training to cover differences in human faces like gender, ethnic origin, illumination, and individual appearance (see section 6.6.2.1) which are not linked to the class specific properties. Thus up to four times as many images need to be annotated manually. Furthermore, the time-consuming AAM analysis has to be performed four time for one image. Regarding the classification performance, it suggests itself that for a proper classification performance, the aimed classes need to show dramatic differences in the appearance such as a head turned to the left versus a head turned to the right. For the problems of gender, facial expression, or even age recognition this method is strongly expected to fail.

Consequently, due to the intractable annotation and computational effort in combination with the expectedly poor classification performance this approach was widely neglected. However, since the annotation effort did not increase due to vertical image mirroring, we successfully applied this procedure to a pre-recognition of the side to which a head might be turned (see section 6.6 Head Pose Recognition).

6.1.2 Statistical classification based on AAM coefficients

As described in section 4.2.3, the coefficient vector $\boldsymbol{h}_c$ in conjunction with the Appearance Model explains the appearance of an unknown object to a large extent, supposing a sufficiently accurate re-synthesis. Consequently the information on numerous facial properties is encoded within $\boldsymbol{h}_c$. In Pattern Recognition language Active Appearance Model analysis can be considered as a feature extraction algorithm. The "decoding" of the information in the coefficient vector shall be performed by a classifier which assigns one *label* out a pre-defined set of labels or classes to the object on hand. For this scenario a static and statistical classifier is the choice. In a training phase a set of images with manually assigned classes is analyzed by an AAM and an edited version of the coefficient vector $\boldsymbol{h}_c$ per image is presented to the statistical classifier.

Thereby, the editing of $\boldsymbol{h}_c$ comprises basically two post-processing steps. On the one hand the coefficients can be represented by their absolute values or normalized to the basis vector specific variance.

$$\overline{\overline{h_{c,i}}} = \frac{h_{c,i}}{\sigma_{c,i}^2} \tag{6.1}$$

Here $\sigma_{c,i}^2$ denotes the variance of the training data in the direction of the ith basis vector (see section 5.2.2.4) which corresponds to the ith Eigenvalue $\lambda_{c,i}$ for PCA-AAMs. On the other hand we can reconstruct the information about the pure shape from $\boldsymbol{h}_c$ applying equation 4.25 repeated here:

$$\boldsymbol{s}(\boldsymbol{h}_c) = \overline{\boldsymbol{s}} + \boldsymbol{Q}_s \boldsymbol{h}_c \, , \, \boldsymbol{Q}_s = \boldsymbol{\Phi}_s \boldsymbol{K}^{-1} \boldsymbol{\Phi}_{cs} \tag{6.2}$$

The shape in $\boldsymbol{s}(\boldsymbol{h}_c)$ is originally represented in Cartesian coordinates. Additionally, the landmark positions of the shape can be transformed to polar coordinates for a better modeling of the variations especially in facial expressions and head pose [7].

The statistical data modeling and optimization of decision instructions for classification is an own area of research. The target is to automatically transform and/or describe the feature spaces created from preferably few training samples and find ways to generalize from the training data to arbitrary unknown data so that the classification performance is accordingly high. For many applications the approach known as Support Vector Machines proved to be best method for classification of static "signals" [101]. In the context of this work Kriegel [63] has shown that this fact is also valid for features space of AAM coefficients.

6.1.3 Support Vector Machines

Support Vector Machines (*abbrev.* 'SVM') as introduced by Vapnik, 1995 [114] and discussed by Burges in [14] are linear and in the general case non-linear classifiers that try to separate positive and negative examples of a training set by a hyperplane. Apart from their classification performance they are also renowned for the low computational costs of the classification, while the training can be comparably lengthy [18].

Let the AM coefficient vector $\boldsymbol{h}_c$ be a test example and $\boldsymbol{h}_{c,i}$ the ith training example with $0 \leqslant i < k$. The classes are labeled with $y \in \{-1; 1\}$ where y_i is the class label of the ith training example respectively. Finally $K(\boldsymbol{h}_c, \boldsymbol{h}_{c,i})$ is a kernel function and $\boldsymbol{\gamma} = (\gamma_0, \ldots, \gamma_{k-1})^T$ the model parameters [84]. The SVM training algorithm minimizes the objective function

$$Q(\boldsymbol{\gamma}) = -\sum_{i=0}^{k-1} \gamma_i + \frac{1}{2} \sum_{i=0}^{k-1} \sum_{j=0}^{k-1} \gamma_i y_i \gamma_j y_j K(\boldsymbol{h}_{c,i}, \boldsymbol{h}_{c,j}) \qquad (6.3)$$

in respect to $\boldsymbol{\gamma}$ preserving the constraints

$$\sum_{i=0}^{k-1} \gamma_i y_i = 0 \qquad \text{with } 0 \leqslant \gamma_i \leqslant C \qquad (6.4)$$

C denotes the *Complexity* of the SVM.

For classification the following decision function has to be solved:

$$y = \text{sign}\left(\sum_{i=0}^{k-1} \gamma_i y_i K(\boldsymbol{h}_c, \boldsymbol{h}_{c,i}) + b\right) \qquad (6.5)$$

b represents the *bias* or offset of the separating hyperplane.

For classification two different kernel functions are most common and provided by most implementations of Support Vector Machines [121, 18]. Firstly, the Radial Basis Function (*abbrev.* 'RBF') kernel

$$K_G(\boldsymbol{h}_{c,i}, \boldsymbol{h}_{c,j}) = e^{-\rho^2 ||\boldsymbol{h}_{c,i} - \boldsymbol{h}_{c,j}||^2} \qquad \text{with } \rho \in \mathbb{R} \qquad (6.6)$$

or secondly a homogeneous polynomial kernel:

$$K_P(\boldsymbol{h}_{c,i}, \boldsymbol{h}_{c,j}) = (\boldsymbol{h}_{c,i} \cdot \boldsymbol{h}_{c,j})^d \qquad (6.7)$$

Both kernels are used to project a linearly non-separable dataset into a higher, linearly separable domain, essentially by splitting the training dataset nonlinearly. The application of the optimal kernel function and its parameters should be determined evaluatively. Throughout this work the polynomial kernel showed best performance and is therefore the basis of all evaluation results except stated differently.

6.1.4 N-fold Cross-Validation

It might have caught the attention that SMVs are designed for two-class problems only. When it comes to tasks dealing with multiple classes, two approaches are commonly implemented [101]: The Multi-layer SVM constitute a kind of decision tree with single SVMs at the junctions. For the mentioned example of our four class facial expression problem this could read like fusing the classes (*neutral, frown*) and (*yell, smile*) for distinguishing these on the top layer. The final decision would be drawn by SVMs which then separate between *neutral* and *frown* or *yell* and *smile* respectively on the second layer.

The second and even more spread approach is the creation of "one-against-all" SVMs. This means that for the mentioned case four different SVMs are trained, e.g. *smile* as class $y_1 = 1$ and (*neutral, frown, yell*) as class $y_0 = -1$. For this purpose the term of equation 6.5 is computed providing the orientation and distance of the test sample h_c to the separating hyperplane which can be interpreted as confidence. Finally, the SVM with the highest confidence determines the class decision.

Many implementations are freely available, that implement Support Vector Machines. We use the machine learning platform WEKA for our evaluation and the open source library Torch3 for online classification integrated to the FacE Analysis System FEASy (see 1.2) [121, 18, 45].

6.1.4 N-fold Cross-Validation

Face analysis, just like most pattern recognition problems, suffers from data sparseness, i.e. the situation that the number of available test samples (images) is low compared to the dimensionality of the features spaces. For any evaluation a set of images for classifier training and a disjunctive set for testing is required. Thereby, both sets should be *representative*, i.e. they should individually cover the whole range of characteristics which can be observed in real life scenarios. However, even large image databases consisting of some thousand images can hardly meet this requirement, even less when they are divided in test and training sets. Thus an evaluation of pattern recognition systems, mainly assembled by feature extraction and classification modules, can only be an estimation of the system capability of generalization from a training set to an arbitrary test set as it would be necessary for real life applications.

In order to overcome the data sparseness and still provide a sophisticated estimation of the system performance, the methodology of n-fold cross-validation is applied [62]. This means that the dataset is split in n subsamples. Depending on the problem these n datasets should be disjunctive. For instance at all tasks, except person identification, all sets should be person disjunctive, i.e. the images of one person appears exclusively in one of the n datasets. Again this simulates the real application scenario, where most likely no previously known faces will be analyzed.

From the n subsamples, a single subsample is retained as validation data for testing the classification and the remaining $(n-1)$ subsamples are used as training data. The cross-validation process is then repeated n times, while each of the subsamples is used exactly once as validation data. The n results from the folds can be combined, e.g. averaged, to give a single estimation after all.

6.2 Image Databases

In order to get adequate databases for the different recognition problems, various face databases were utilized for the AAM generation and the evaluation of the AAM (Active Appearance Model) search algorithm. Before improving an algorithm, it is necessary to ensure that the training data used is optimal. If the training data is not selected carefully, the algorithm might show poor results no matter how good the improvements may be. Therefore, depending on the evaluation problem, a small subset (between 1.5% and 10%) from the AR Database [73], NIFace1 Database [2], FG-NET Aging Database [1] or MMI Face Database [89] is selected for training the AAM. Then the complete database or a large subset of it is used to evaluate the AAM search algorithm. Table 6.1 gives an overview over the attributes and (subjective) quality of images from the various face databases introduced in this chapter. All databases except for the FG-NET Aging Database consist only of color images.

Database name	Dimensions	Format	Subjective quality
AR Database	768×576	BMP	good
NIFace1 Database	640×480	PNG	medium
FG-NET Aging Database	avg. 400×500	JPEG	bad, partly black/white
MMI Face Database	1200×1600	JPEG	very good

Table 6.1: Comparison of face databases

6.2.1 The AR Database

The AR Database [73] which was created by Aleix Martinez and Robert Benavente in the Computer Vision Center (CVC) at the U.A.B (Universitat Autònoma de Barcelona) includes 3315 pictures of 135 people. The images feature frontal view faces with different facial expressions, illumination conditions, and occlusions (sun glasses and scarf). Each person participated in two sessions, separated by two weeks (14 days) time. The same pictures were taken in both sessions.

For the classification of gender, facial expression and face identification in this work only a subset of the AR Database is used. This subset consists of pictures which feature neither different illumination conditions nor occlusions (sun glasses and scarf). This results in a set of 1020 pictures (564 male, 456 female) including up to 8 pictures per person with different facial expressions (neutral, smile, anger, scream) taken in the two sessions. An example of the 4 facial expressions is shown in figure 6.1.

6.2.2 The NIFace1 Database

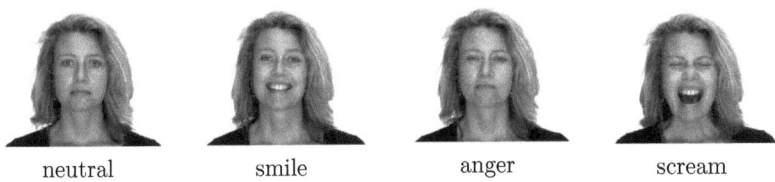

neutral	smile	anger	scream

Figure 6.1: Example of different expressions from the AR Database

6.2.2 The NIFace1 Database

The NIFace1 Database [2] was created at the Faculty of Computer Science and Automation of the Ilmenau University of Technology. This database includes 2610 images of 90 people with neutral facial expression, different illuminations and small deviations in head pose. Here head pose refers to the direction of head orientation. Identities are equally distributed in the age range between 10 and 60 and equally distributed over genders. The 15 different poses as shown in figure 6.2 are composed of 5 different horizontal angles $\alpha_H \in \{30°, 15°, 0°, -15°, -30°\}$

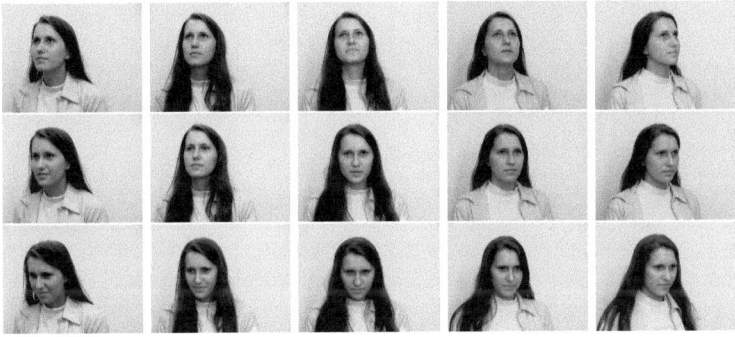

Figure 6.2: Example of different head poses from the NIFace1 Database

from left to right and 3 different vertical angles $\alpha_V \in \{20°, 0°, -20°\}$ from bottom to top. This database is used for classification of head pose.

6.2.3 The FG-NET Aging Database

The FG-NET Aging Database [1] contains 1002 pictures of 82 persons at different ages. It has been generated as part of the European Union project FG-NET (Face and Gesture Recognition Research Network). Figure 6.3 shows example images of different persons at different ages. These images feature faces with different head poses, illuminations, facial expressions, glasses and hats. The age histogram ranging from 0 to 69 years is shown in figure 6.4. The average age of the FG-NET Aging Database is 15.84 (the median is 13). This means that the database consists of a lot of images with people at young age.

| 0 | 6 | 21 | 31 | 43 |

Figure 6.3: Example of different ages from the FG-NET Aging Database

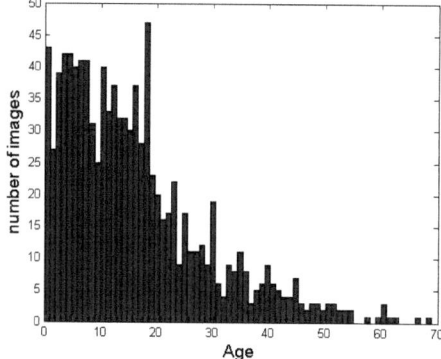

Figure 6.4: histogram of the age variation

The database includes black and white as well as color images. If you train an AAM on both of these image types, the performance will be suboptimal due to the fact that the AAM will extract a lot of variation on the color. This information is not important for the age classification. Therefore we only used the 811 color images when evaluating the age of a person.

The FG-NET Aging Database already contains annotations (process of placing landmarks on a face image) of all images. An example of the annotations with 68 landmarks is shown in figure 6.5.

6.2.4 The MMI Face Database

The MMI Face Database [89] was created at the Delft University of Technology in the Man-Machine Interaction Group. This database includes 740 images of 19 persons with varying facial expressions and poses (frontal and profile view). We only applied the MMI Face Database for training the AAM which was used for the gender, facial expression and face identification classification.

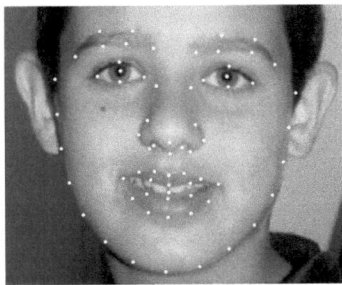

Figure 6.5: Example of the FG-NET Aging annotations with 68 landmarks

6.3 Gender Recognition

6.3.1 Dataset

The evaluation of gender recognition was run on a subset of the AR Database [73], as described in section 6.2.1.

The gender evaluation was performed using an AAM generated from a set of 50 images from the AR Database and 9 images from the MMI Face Database [89] together with 78 landmarks (see [32]). This set includes images from 10 male persons (25 images) and 13 female persons (34 images).

6.3.2 Results

The subset of 1012 pictures (557 male, 455 female) is evaluated using FEASy. We evaluated the 2-class problem using the Support Vector Machines (*abbrev.* 'SVM') [14] as classifier, performing a 10-fold cross validation. Setting the exponent for the polynomial kernel to 6, we received a recognition rate of 94.6% on AAM coefficients extracted by the offline optimization approach. Recognition rate in this work is defined as the percentage of correctly classified objects to total objects. We also performed the same evaluation on only 72 landmarks as introduced in figure 4.2(a) and the best result was a recognition rate of 92.2% for SVM with an exponent of 3. The six additional landmarks were introduced to obtain a better symmetry of the shape model. They were placed on the forehead and on back of the nose. This shows that the additional six landmarks improve the result by about 2.2%.

However when using the 10-fold cross validation implemented in WEKA [121], there is no guarantee that training set and dataset of one single fold do not both include pictures from the same person. If this was the case, a person from the dataset would already be known and thus a better recognition rate is possible. Therefore we implemented a 5-fold cross validation where the 5 datasets were person-disjunctive. Each dataset contained about the same amount of male and female pictures. A 5-fold cross validation means that 80% of the faces are used for training and 20% are used for testing. With the SVM and an exponent of

2 the recognition rate resulted in 87.5%. The confusion matrices and evaluation results for both classifications are summarized in table 6.2. Class M refers to male, F to female subjects.

	M	F
M	525	32
F	23	432
Total error:	5.4%	

(a)

	M	F
M	485	72
F	54	401
Total error:	12.5%	

(b)

Table 6.2: Cross-classification confusion matrices and error rates for gender evaluation; (a) non person-disjunctive and (b) person-disjunctive

The confusion matrices show that in both cases male faces are classified as female more often than female faces as male. One reason is that this dataset contains more pictures of male faces. But also the ratio of incorrectly classified men to women is higher than the actual ratio of men to women in the dataset. This indicates that women are classified correctly more often. For the person-disjunctive case, the recognition rate for women is 88.1% and about 1% higher than the recognition rate for men. For the non person-disjunctive evaluation the recognition rate for women is only about 0.7% higher.

6.3.2.1 Comparison of Online Optimization Strategies

For the 5-fold, person disjunctive dataset we performed an evaluation applying the online optimization techniques. Table 6.3 shows the results.

Method	# Face syntheses	Texture similarity	Recognition rates
Offline	28.3	.928	87.5%
Gradient descent	3750	.904	85.1%
Montecarlo sampling	3110	.954	92.4%
Simplex	299	.963	93.2%

Table 6.3: Comparison of different optimization methods

The results show clearly the correlation between the texture similarity and the recognition rate. Furthermore, the Simplex algorithm showed the best results regarding re-synthesis and classification. Additional evaluations [84] confirmed this observation. Since the Simplex is characterized by a multiple higher efficiency, it is used as benchmark for all further evaluations.

6.3.2.2 Selection of Images with Best Representation

Now we perform the same person-disjunctive 5-fold cross validation sorting out images with bad AAM representation. To evaluate which input images can be analyzed best we use the Texture Similarity (*abbrev.* 'TS') measure suggested in equation 5.96. In figure 6.6 top, a histogram of the TS values for the gender problem is shown.

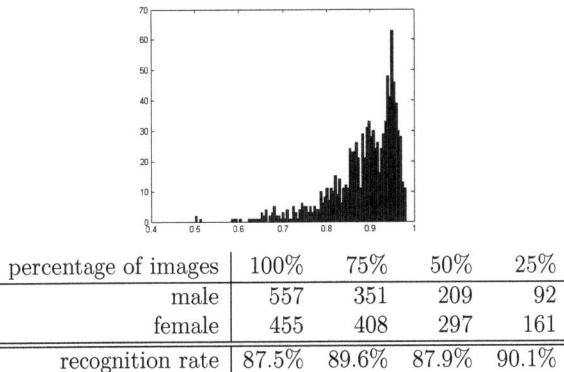

percentage of images	100%	75%	50%	25%
male	557	351	209	92
female	455	408	297	161
recognition rate	87.5%	89.6%	87.9%	90.1%

Figure 6.6: Texture Similarity histogram and evaluation results

When choosing images with a TS larger than 0.85, 0.90 or 0.94, we received a corresponding subset consisting of the best represented 75%, 50% or 25% of the images. Figure 6.6 bottom summarizes the amount of male and female images and the resulting recognition rates for the original 5-fold cross validation and these 3 database reductions. The histogram shows that most of the images have a TS value closer to 1 than to the minimum of 0.5. Therefore choosing the 75% images represented best by the AAM with a TS value larger than 0.85, improves the recognition rate by about 2%. Since the TS values of the remaining images are not widely spread, the recognition rate is not significantly improved any further. For the best 50% of the images, the recognition rate even worsens.

Also the amount of pictures decreases relatively less for female pictures when decreasing the percentage. In other words female pictures can be represented better. The reason for this could be that male faces have more variations like beards and glasses (more men than women wear glasses) and also women faces are more consistent due to makeup.

6.3.3 State-of-the-Art

There are several other studies which tried to solve the gender recognition problem. Table 6.4 shows our recognition rate in comparison to state-of-the-art recognition rates of different gender recognition systems. None of these studies talks about a person-disjunctive evaluation although all the databases used contain several images per person. So we can use our non person-disjunctive result for comparison. Additionally our other result is also stated in the table.

Kembhavi's [59] experiment, which is based on Support Faces [82], yielded a recognition rate of 96.66% using a subset of 240 images (120 male and 120 female) from the AR Database. Buchala [12] used a subset of the FERET Database [93].

study	face databases	recognition rates
Shunting inhibitory convolution neural networks 2006 [112]	FERET	97.2%
	BioID	86.4%
Support faces 2005 [59]	AR	96.66%
Support faces 2002 [82]	FERET	96.62%
FEASy offline	*AR*	*94.6%*
AAM MLP 2005 [120]	NIFace2 (own)	~ 93%
FaceIt 2001 [48]	FERET	~ 90.6%
	AR	~ 87.5%
PCA-LDA 2005 [12]	FERET	86.43%

Table 6.4: Comparison of our result with state-of-the-art gender recognition rates

Gross [48] compared the recognition rates for men and women and also detected a higher recognition rate for women for the AR as well as for the FERET database. In his study, he only states the results for men and women individually. We took the mean of both results. The recognition rate for the FERET database is about 3% higher than for the AR Database which indicates that the gender recognition for FERET has a better classification. Wilhelm [120] achieved the best result with a Multi Layer Perceptron classifier which was still worse than our result. The newest study [112] based on neural networks has a result for FERET which is almost 3% better than our result but for the BioID database the result is clearly below ours. Overall our gender recognition system is highly comparable or superior to current gender recognition systems.

6.4 Facial Expression Recognition

6.4.1 Dataset

For evaluating the facial expression of a person, a subset of the AR Database [73] was used, as described in section 6.2.1. This subset included 1020 images with the facial expressions neutral, smile, anger and scream. During this evaluation, class N refers to neutral, S to smile, A to anger, Y to yell (or scream). These expressions are shown in figure 6.1.

The facial expression evaluation was performed using the same AAM as for the gender problem generated from a set of 50 images from the AR Database and 9 images from the MMI Face Database [89].

6.4.2 Results

Again we performed the facial expression classification using both a 10-fold cross validation (implemented in WEKA [121]) and the same person-disjunctive 5-fold cross validation provided for the gender problem (see section 6.3.2) on an SVM with an exponent E of 1 and a complexity parameter C of 1. The confusion matrices and evaluation results are summarized in table 6.5.

6.4.2 Results

	N	S	A	Y
N	182	6	60	4
S	12	230	7	5
A	83	3	164	5
Y	3	12	5	231
Total error:				**20.3%**

(a)

	N	S	A	Y
N	180	11	56	5
S	12	229	4	9
A	89	4	155	7
Y	5	12	8	226
Total error:				**21.9%**

(b)

Table 6.5: Cross-classification confusion matrices and error rates for facial expression evaluation; (a) non person-disjunctive and (b) person-disjunctive

Contrary to the gender problem the recognition rates for the non person-disjunctive and the person-disjunctive case are very similar here. This indicates that for expression classification of a person it does not make a big difference if the person is already known to the training data. The reason for this is that here class variance (different facial expressions) is already included within one person. The confusion matrices mainly show that the classifier has difficulties to distinguish between the angry and neutral expression. Both expressions differ in very few aspects (see figure 6.7) and therefore the coefficient vector space of these 2 classes of the AAM falls into the same region.

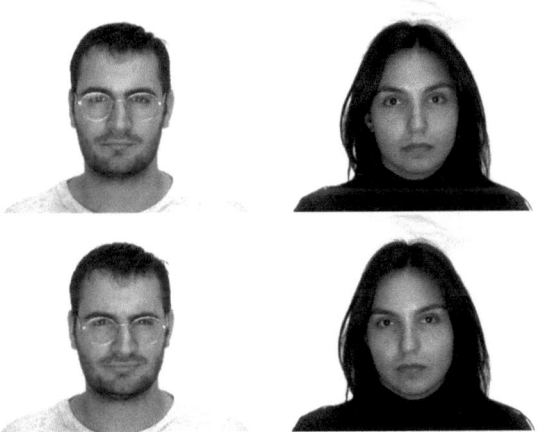

Figure 6.7: Comparison of neutral (top) and anger (bottom) expression

We conducted a survey where 15 persons were required to label all 1020 database images according to the facial expression. The average recognition rate was poor 84.6%. In the light of this human classification performance, the results of the automatic analysis are even more satisfying.

Therefore a good classification is difficult. When reducing this 4-class problem to a 3-class problem by leaving out images of class A (anger), the recognition rate for the non person-disjunctive case results in 93.3%. When applying the Simplex optimization, we obtained 81.9 % recognition rate for the 4-class problem and very good 95.1% for the three-class problem.

The recognition rates could not be improved any further by selecting only images with good AAM representation.

6.4.3 State-of-the-Art

Again, when comparing our result to results of other studies none of these talk about a person-disjunctive evaluation although all the databases used contain several images per person. Therefore, table 6.6 shows both of our non person-disjunctive recognition rates in comparison to state-of-the-art recognition rates of different facial expression classification systems.

study	face databases	recognition rates
FEASy Simplex, 3 classes	AR	95.1%
FEASy Offline, 3 classes	AR	93.3%
line-based Caricatures 2003 [44]	AR	86.6%
Cootes AAM [7]	AR	84.4%
single training sample PCA 2003 [72]	AR	$\sim 84\%$
FEASy Simplex, 4 classes	AR	81.9%
Cascaded AAM 2006 [96]	AR	79.9%
FEASy offline, 4 classes	AR	79.7%
ICA NN 2005 [120]	NIFace2 (own)	$\sim 73\%$

Table 6.6: Comparison of our result with state-of-the-art facial expression recognition rates

All studies used the AR Database except for Wilhelm's AAM MLP [120]. But his classification was performed on 8 different facial expressions. Saatci's [96] Cascaded AAM result for a 4-class expression classification with a preceding gender recognition is equal to our 4-class result. He first classified the gender of a person and then performed the facial expression recognition separately for each gender. This improved the performance of his system by 3.5% (from 76.4%) whereas using this approach on our system lead to a worse result. Diduch [31] who performed similar evaluations only used a very small subset of the AR Database (200images) and therefore his result is better but less representative. Also Basili [7] only used 384 images. Martinez [72] and Gao [44] both implemented an expression evaluation of 3 classes by discarding one of the classes which is difficult to recognize (neutral and anger). Their results were significantly worse than our 3-class recognition rate.

6.5 Person identification

6.5.1 Dataset

For face identification, a subset of the AR Database [73], as described in section 6.2.1, was used. However only persons who participated in both sessions (14 days apart) were relevant for this face identification. An example of images taken from a person at both sessions is shown in figure 6.8.

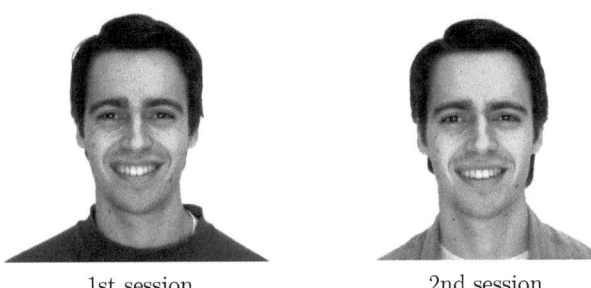

1st session   2nd session

Figure 6.8: Example of images (smile) taken 14 days apart

The first session of these 120 persons formed the training set (478 images) whereas the images from the second session were used as dataset (475 images). This setup means that the face of a person should be recognized on a training set consisting of images taken 14 days earlier.

The face identification was performed using the same AAM as for the gender and facial expression problem generated from a set of 50 images from the AR Database and 9 images from the MMI Face Database [89].

6.5.2 Results

The face identification was performed applying K* [17], an instance-based classifier (implemented in WEKA [121]), with a global blending parameter of 95 to the training set.
The recognition rate for the face identification of the 120 subjects resulted in 58.3%. However we reduced the dataset to 100 subjects by eliminating the 20 subjects with worst face identification. Now 67.9% of the faces were classified correctly. These results are summarized in table 6.7.

number of subjects	120	100
recognition rate	58.3%	67.9%

Table 6.7: Results for the face identification

6.5.3 State-of-the-Art

Table 6.8 shows our face identification rate in comparison to state-of-the-art recognition rates of different face identification systems.

study	face databases	recognition rates
SOM-Wavelet Networks HMM 2005 [127]	ORL	91.36%
HE+LN Algorithm 2005 [125]	AR	87.1%
FEASy offline	AR	*67.9%*

Table 6.8: Comparison of our result with state-of-the-art face identification rates

Yang Zhi used SOM-Wavelet Networks [127] on 50 persons from the ORL Database [97] whereas Xie [125] implemented an algorithm to recognize 121 subjects from the AR Database. These classifications were performed on images taken at the same session. However our system performance for face identification on 100 persons is poor compared to these other systems.

6.6 Head Pose Recognition

6.6.1 Dataset

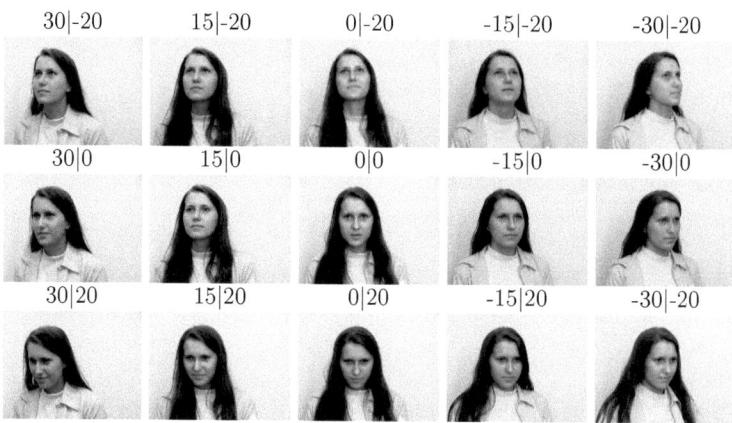

Figure 6.9: Different classes for the head pose evaluation

For evaluating the head pose of a person, the NIFace1 Database [2] (see section 6.2.2) was used. When utilizing FEASy the AAM search requires both eye positions. The NIFace1 Database contains only head pose angles of a maximum of 30 degrees and therefore both eye positions are still mostly visible. The ET (Head and Eye-Tracking Module) only had problems to find the ROI information of about 100 images from the complete database. In figure 6.9 the classes used in this work are listed above the corresponding image. The first number refers to the horizontal angle whereas the second number after the "|" refers to the vertical angle of the head pose.

6.6.2 Results

All the evaluations for the head pose recognition were performed using a 10-fold cross validation on an SVM with an exponent E of 1 and a complexity parameter C of 1 implemented in WEKA [121] if not stated differently.

6.6.2.1 Image Selection for AAM Generation

When selecting the training AAM for the head pose evaluation, we only considered images from the left side. With left side we refer to images with a horizontal angle $\alpha_H \in \{30°, 15°, 0°\}$ represented by the left 9 poses in figure 6.9. Here the optimization of the left side AAM is performed whereas in subsection 6.6.2.3 a method is introduced how we can apply this model to all head poses. The set of images annotated for the different AAMs always included an equal number of images of all 9 left positions and 2 additional images from the AR database. Only pictures with an artificial illumination were selected because these pictures were generally more focused.

In some of the images the points where the landmarks would usually be set are not visible due to the different head poses. These landmarks were set along the texture edge assuming where the actual positions of these points are. Figure 6.10 shows an example of an annotated image of class 30|-20. The rightmost point of the the left eye (corresponding to landmark 33) is covered by the nose and therefore this landmark is set along the nose assuming where this point actually is.

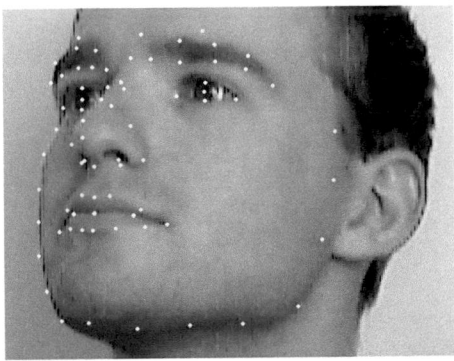

Figure 6.10: Example of an annotated image from the NIFace1 Database

The different AAMs used to evaluate all images with artificial illumination from the left side consisted of 20, 38, 47, 56 and 65 images. An AAM for training should be made from images with lots of variance so they can represent the applied dataset well. The images we used for the model contained variations like beards, glasses and poses. After choosing 20 images with different variations for the first model, we performed an evaluation on this model. Then we selected images with a bad TS value for the next 18 images which we added to the model. This procedure was repeated to obtain the additional images we added for the models with 47, 56 and 65 images. The mean TS values and the recognition rates for these models after AAM search are listed in table 6.9.

number of images	20	38	47	56	65
mean TS	0.769	0.820	0.832	0.858	0.850
recognition rate	52.0%	58.8%	58.1%	60.2%	58.9%

Table 6.9: TS values and recognition rates for different amounts of images for AAM training

The AAM consisting of 56 images gave the best results for mean TS and recognition rate.

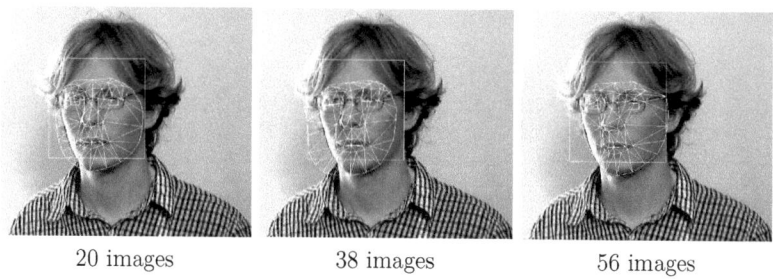

20 images   38 images   56 images

Figure 6.11: Improvement of the AAM during AAM Search

Figure 6.11 shows how the AAM improved during AAM Search for the models with 20, 38 and 56 images. The AAM with 56 images converges pretty well to the actual face shape during the AAM search. A further increase of the number of model images resulted in a worse evaluation performance. This leads to the assumption that the size of approximately 60 images is a good size when choosing the AAM because also for the AR database the model consisted of 59 images. The AAM with 56 images will be used for the remaining evaluations if not stated differently. The confusion matrix for classification with this model is listed in table 6.10.

The confusion matrix indicates that the classifier has difficulties to distinguish between the different vertical angles but not so much between the horizontal angles. When reducing the 9-class problem to two 3-class problems by combining the classes (new SVM classification), the recognition rate for horizontal differentiation results in 92.1% but for vertical differentiation it is only 75.0%.

The confusion matrices for these 3-class problems are listed in figure 6.11. The confusion matrices indicate that our system has problems to distinguish between

6.6.2 Results

	0|20	0|0	0|-20	15|20	15|0	15|-20	30|20	30|0	30|-20
0|20	55	21	2	5	2	0	0	0	0
0|0	18	43	20	0	3	1	0	0	0
0|-20	2	17	58	0	3	4	0	0	1
15|20	4	1	0	55	21	1	2	0	0
15|0	1	1	1	12	56	11	2	0	1
15|-20	0	1	2	2	22	50	1	1	4
30|20	0	0	1	3	4	0	54	15	4
30|0	0	0	0	1	5	2	16	37	20
30|-20	0	0	0	1	2	4	4	21	36

Total error: 39.8%

Table 6.10: Cross-classification confusion matrix and error rate for head pose evaluation with a 56 image AAM

	0	15	30
0	484	36	1
15	19	470	27
30	1	36	451

Total error: 7.9%

(a)

	20	0	-20
20	403	97	9
0	78	350	89
-20	16	92	391

Total error: 25.0%

(b)

Table 6.11: Cross-classification confusion matrices and error rates for the 3-class head pose problem; (a) horizontal angles and (b) vertical angles

images with different vertical head poses. The reason for this is that the AAM does not change as much for vertical as it does for horizontal head movement. Fig 6.12 shows the AAMs for the 3 different horizontal and vertical head poses.

6.6.2.2 Parameter Optimization

The AAM trained on 56 images will now be optimized by varying different parameters.

Combined Percentage Parameter Our first approach suggests to reduce the combined percentage parameter of the PCA. This parameter controls the number of Eigenvectors selected for the Shape and Texture PCA for the final AAM as described in chapter 4. Only the combined Eigenvectors whose corresponding Eigenvalue's cumulative sum is below this percentage of the total sum are used. These Eigenvalues are called principal components.

Figure 6.13 shows that the Eigenvector corresponding to the first principal component of our head pose AAM represents the horizontal head movement (σ is the standard deviation). The second Eigenvector describes the vertical head movement. Therefore, the first 2 principal components already include the main

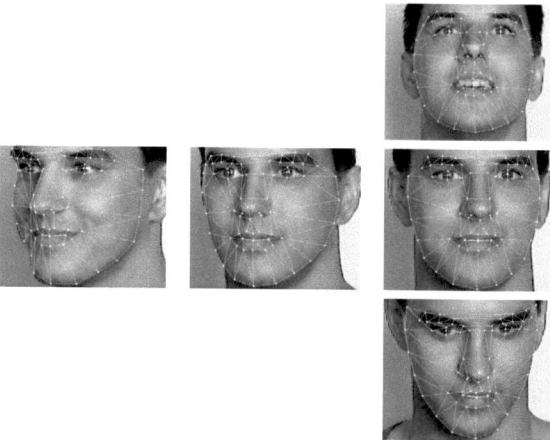

Figure 6.12: AAMs for different head poses

information for the head pose problem. Now we assumed that decreasing the number of principal components would lead to a better result. The results for the reduction of the default combined percentage parameter (0.98) are listed in table 6.12. The recognition rate and the mean TS value both decreased and thus the combined percentage parameter remained 0.98 for the additional head pose evaluations.

combined percentage	98%	95%	90%
mean TS	0.858	0.845	0.828
recognition rate	60.2%	57.7%	55.1%

Table 6.12: TS values and recognition rates for different combined percentages

Combined Model and Euclidean Parameters Our next approach was to vary the combined model parameter deviation p and the Euclidean parameters t_x, t_y, s and r (introduced in chapter 4). The AAM is trained to vary these parameters according to their selected value during AAM Search. Due to the fact that our image database includes various head poses, allowing the AAM Search to perform more translation, scaling, rotation and deviation should give a better result. Figure 6.14 shows the mean TS values and the recognition rates for varying these 5 parameters.

The default parameter values form the starting point of the figure. These are then individually multiplied by a Parameter Factor which is listed along the x-axis. The results for these different AAMs were evaluated and are listed along the y-axis. Multiplying the scaling parameter s by 4 ($s = 0.4$) leads to the best recognition rate of 65.4% and a mean TS value of 0.870. When further increasing s, the mean TS also ligthly increases although the recognition rate worsens. This

6.6.2 Results

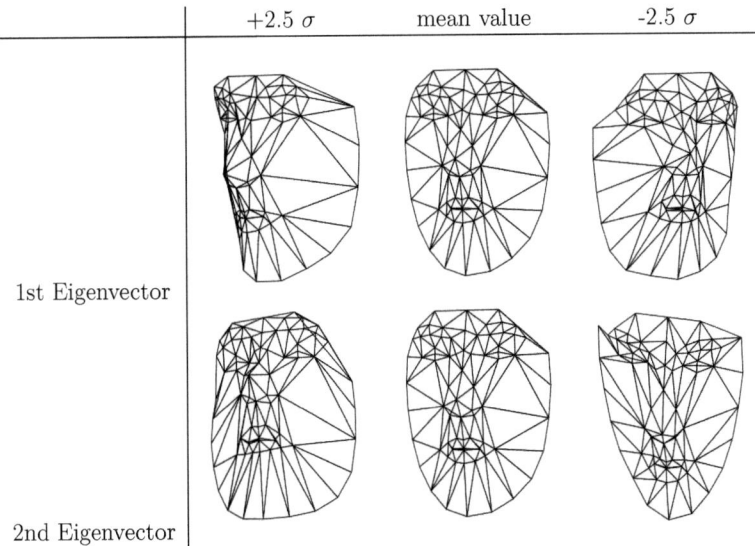

Figure 6.13: Effect of the first 2 AAM principal components

indicates that we do not always have a dependency between the mean TS value and the recognition rate.

When increasing t_y, both results quickly decline. In this case the search area is allowed to query areas outside the picture.

So far, we only incremented these 5 parameters separately. But now we want to combine setting s to 0.4 (the best result from above) with individually adjusting each of the 4 remaining parameters to the values where we received the best individual recognition rate. This means multiplying the default value of t_x by 2, t_y by 3, r by 2 and p by 3. The results for these 4 combinations (in boldface) are compared to our best result for the individual parameter optimization in table 6.13. The last row shows our result before parameter optimization.

s	t_x	t_y	r	p	mean TS	recognition rate
0.4	0.1	0.15	0.03	2.0	0.870	65.4%
0.4	**0.2**	0.15	0.03	2.0	0.870	63.9%
0.4	0.1	**0.45**	0.03	2.0	0.861	63.4%
0.4	0.1	0.15	**0.06**	2.0	0.869	62.8%
0.4	0.1	0.15	0.03	**6.0**	0.875	64.2%
0.1	0.1	0.15	0.03	2.0	0.858	60.2%

Table 6.13: TS values and recognition rates for different parameter combinations

The recognition rate was worse for these 4 combinations than for only optimizing the scaling parameter s. This indicates that these 5 parameters depend on each other. The recognition rate could not be improved anymore by further optimizing these parameters.

These evaluations were only performed on images with artificial illumination. When applying the AAM with $s = 0.4$ and the other parameters set to default values on the images with artificial and natural illumination, the recognition rate results in 67.1% for the 9-class problem. Using an exponent of 2, this result could be optimized to 68.9%.

6.6.2.3 Different Systems for the 15-Class Problem

Until now, the system only classified different poses from the left side. Therefore we want to classify the 15-class problem of all poses with 2 different systems. The first system basically uses an AAM including all 15 pose variations, the second applies our left side AAM twice on all images and on all images flipped horizontally (along the vertical median axis of the image) and selects the image with better convergence.

15-Class System   For the first system we used the annotations performed by Störmer [109] [110] which consist of 132 landmarks as shown in figure 6.15. First we trained an AAM on 45 images ($s = 0.4$) selected from the left side.

These images were chosen according to the criteria mentioned in subsection 6.6.2.1. The recognition rate on all images from the left side amounted to 66.4% (with C=3) which is only 2.4% less than for our AAM with 56 images (annotated with fn78 landmarks). Additional 30 images with $\alpha_H \in \{-15°, -30°\}$ were added to this model, to form an AAM consisting of 75 images including all 15 different pose variations. When classifying all images from the NIFace1 database on this AAM, the recognition rate for this problem resulted in 60.5% (complexity parameter of 3).

9-Class Mirrored System   For our second system we applied the left side 56 images AAM ($s = 0.4$) twice. First we applied it to all images from the image database although the AAM can only represent head poses where $\alpha_H \in \{-15°, -30°\}$ rather badly. Next we flipped all images in horizontal direction and again performed the AAM search on all these flopped images. Hereby the images' head orientation to the right were mirrored so that the pose was now towards the left. Due to this horizontal flip, our AAM does not need to model the additional variance of the head poses with angle $\alpha_H \in \{-15°, -30°\}$.

Side Pre-selection   We implemented a side pre-selection by comparing the TS value for the original and flipped image and selecting the image with the larger TS value. This pre-selection decides, if an image has a left or right head orientation. If the TS of the original image was higher, then we would classify it as image from the left side, otherwise as image from the right side. For images with $\alpha_H = 0°$ it did not matter which TS was higher. The total error rate for this pre-selection amounted to 1.1%.

6.6.2 Results

Then the classification was performed on the 9 different head positions of the left model which together with the pre-selection formed the 15-class problem. Taking the pre-selection error into account, the recognition rate was 68.1% with a complexity parameter of 3 and only 0.8% lower than for the 9-class problem with the same AAM. This indicates that images with angles $\alpha_H \in \{-15°, -30°\}$ are more accurate which was verified by a subjective examination of the images. Table 6.14 summarizes the results for the 2 different systems. The 9-class mirrored system definitely is the better approach to solve the head pose problem. The recognition rate for evaluating the 5 horizontal head poses resulted in 89.8% and for the 3 vertical poses in 75.6%. If the system has to distinguish between the poses left, center and right the recognition rate amounts to 94.8%.

	15-Class	9-Class Mirrored
recognition rate	60.5%	68.1%

Table 6.14: Results for different 15-class problem systems

Selection of Images with Best Representation Again, we select different subsets of images which can be represented well by the AAM. The recognition rates for choosing the best 75%, 50% and 25% images are summarized in table 6.15. The number of images with $\alpha_H \in \{15°, -15°\}$ and $\alpha_V = 0°$ was significantly higher than for other head poses when reducing the amount of images.

percentage of images	100%	75%	50%	25%
amount of images	2468	1851	1234	617
recognition rate	68.1%	69.5%	72.1%	72.3%

Table 6.15: Evaluation results for images with best representation

These 2 poses (normal and mirrored) represent the center of the left AAM. The result shows that a better recognition rate is obtained by choosing images with good AAM representation.

6.6.2.4 Manual ROI Input of AAM search

The ET (Head- and Eye-Tracking Module) has problems to recognize the ROI (region of interest) when applying images with different head poses. Thus we want to evaluate if a manual input of the ROI information would also improve our results. Hence, FEASy was initialized without the ET. The AAM search still needs the information about the ROI to perform an AAM search. Therefore the ROI was calculated for images annotated by Störmer [109] [110] and manually put into the AAM re-synthesis. The Störmer annotations where performed for 221 images of whereas 151 are images from the left side.

The ROI information is given by 3 bounding boxes, one for the head (h), the left eye (l) and the right eye (r). Each bounding box is described by 4 parameters: x- (x) and y-coordinate (y) of the upper left point and the height (h) and width (w) of the box. Mathematically, the ROI was approximated from the 132 landmarks $\{(x_0, y_0), \ldots, (x_{131}, y_{131})\}$ of the 221 annotated images by the following formulas (e.g. h_h is the height of the bounding box for the head):

$$h_x = min(x_0, x_8, x_{108}) - 10 \quad (6.8)$$
$$h_y = min(y_{102}, y_{118}) - 30 \quad (6.9)$$
$$h_h = max(y_{12}, y_8, y_{108}) - h_y + 10 \quad (6.10)$$
$$h_w = h_h \quad (6.11)$$
$$l_x = x_{82} - 4 \quad (6.12)$$
$$l_y = y_{82} - 2 \quad (6.13)$$
$$l_h = 8 \quad (6.14)$$
$$l_w = 4 \quad (6.15)$$
$$r_x = x_{99} - 4 \quad (6.16)$$
$$r_y = y_{99} - 2 \quad (6.17)$$
$$r_h = 8 \quad (6.18)$$
$$r_w = 4 \quad (6.19)$$

We used our 45-image and 75-image AAMs mentioned in 6.6.2.3 to classify the 151 left side pose and the 221 total pose pictures. This classification was performed with the ET as well as without the ET and manually providing the AAM search with the ROI values calculated for each image from the equations above. The results are listed in table 6.16.

	Eyetracker	manual ROI input
45-image AAM (left side)	54.3%	58.9%
75-image AAM (all poses)	38.8%	43.8%

Table 6.16: Recognition rates with and without Head and Eye-Tracking Module

For both AAMs the recognition rate was improved significantly. This indicates that the different recognition rates for the head pose problem would be higher for a better performing ET.

6.6.3 State-of-the-Art

It is difficult to compare our head pose recognition rates to the results of other studies. The reason is that there are many different head pose problems with different angles and different amounts of classes. Tables 6.17 and 6.18 compare our results for horizontal and vertical differentiation to the results of other studies.

study	face databases	recognition rates
Coarse Head Pose 2002 [11]	CMU PIE	96%
Geodesic Distance SVM 2006 [70]	CAS-PEAL	90.71%
	FERET	78.79%
FEASy offline	NIFace1	*89.8%*
Neural Networks 2004 [108]	Pointing 04	58.0%

Table 6.17: Comparison of our 5-class horizontal result with other horizontal head pose recognition rates

Brown [11] in her Coarse Head Pose approach achieves a recognition rate of 96% on 5 different horizontal classes [104], where the angles of these classes are $\alpha_H \in \{90°, 45°, 0°, -45°, -90°\}$. Bingpeng [70] applies a Geodesic Distance SVM on two databases, manually labeling the eye positions. For classification 7 classes (same angles as our 5 classes and additional angles of $45°$ and $-45°$) are utilized. The recognition rate applying the CAS-PEAL Database [43] is better than our result which could be improved by also manually inputting the eye positions (see subsection 6.6.2.4).

study	face databases	recognition rates
FEASy offline	NIFace1	*75.6%*
Neural Networks 2004 [108]	Pointing 04	66.0%

Table 6.18: Comparison of our 3-class vertical result with other vertical head pose recognition rates

Stiefelhagen [108] uses the Pointing 04 Database [47] to classify 13 horizontal and 7 vertical angles. However the results stated in the tables are the recognition rates of the same 5 horizontal and 3 vertical ($\pm 30°$ instead of $\pm 20°$) classes as in our problem. The confusion matrices from Stiefelhagen were reduced to receive the stated results.

6.7 Age Recognition

6.7.1 Dataset

For age classification the color images from the FG-NET Aging Database [1] (see section 6.2.3 were used. This work concentrates on 2 different age evaluation problems. First the database is divided into 4 age groups. Class 0 refers to images with persons at the age of 0-19, class 20 to 20-29, class 30 to 30-39 and class 40 to 40-69. The database contains many images of persons at young age. Due to the fact that for a realistic classification the amount of images per class should be about equal, this evaluation could only be performed on 260 images (65 per class) which were randomly selected. Furthermore the database was divided into different 2-class problems separating the database at the age thresholds of 10, 14, 18, 20 and 30.

6.7.2 Results

All the evaluations for the age recognition were performed using a 10-fold cross validation on a SVM with an exponent E of 1 and a complexity parameter C of 1 implemented in WEKA [121].

6.7.2.1 Image Selection for AAM Generation

The FG-NET Aging Database contains many variations like head poses, blurry pictures and different illuminations. Therefore it was important to include all these variations during training of the AAM. The performance of the AAM was verified by applying the 4-class age problem.

Annotations First the annotations provided by the database were manually improved for a subset of 110 images. This means that the positions of the 68 landmarks from an image are corrected if necessary. The results for classifying the 4 age groups on AAMs using the provided and improved annotations are compared in table 6.19. When using manually improved annotations, the recognition rate for the AAM consisting of 100 images was increased by 6.3%.

annotations	FG-NET Aging	manually improved
50 images AAM	36.3%	38.3%
100 images AAM	40.8%	47.1%

Table 6.19: Recognition rates with and without manually improved annotations

Image Amount Selection The manually improved annotations are now used to select the amount of images for the AAM. These results for different amounts of images are listed in table 6.20. When applying the AAM consisting of 100 images to the AAM Search algorithm, the age recognition rate achieved the best

number images per AAM	50	74	92	100	110
mean TS	0.863	0.872	0.883	0.869	0.896
recognition rate	38.3%	35.4%	40.4%	47.1%	42.5%

Table 6.20: TS values and recognition rates for different amounts of images for AAM training

result. Interestingly the mean TS value for this model is worse than for the AAMs (with worse recognition rates) trained on 74, 92 and 110 images. This leads to the assumption that selecting images with good AAM representation is no indicator for a better classification performance. The reason, why the amount of images for the best AAM here is clearly higher than for the AAMs of the other evaluation problems, could be that the FG-NET Aging Database contains more variation. For this setup, blurry images are very important to be included into the AAM.

6.7.2.2 Parameter Optimization

Due to a lot of variation in the database, the same parameter optimizations were performed as for the head pose problem (see subsection 6.6.2.2). Neither the reduction of the combined percentage parameter nor the variations of the combined model parameter deviation p and the Euclidean parameters t_x, t_y, s and r resulted in a better classification.

6.7.2.3 Feature Selection

For evaluating the age of a person, information about rotation, skin color and glasses is not important and might lead to a worse classification result. Some features (here principal components) are supposed to contain this unimportant information. Therefore we performed a Feature Selection (*abbrev.* 'FS') on the 2-class and 4-class problems to improve the performance.

2-Class Problem For this problem the color subset of the database is divided into 2 classes at an age threshold (e.g. 0-19 and 20-69). For the age thresholds 10, 14, 18, 20 and 30 this dataset is reduced so that the number of images per class is equal. The amount of images in these datasets and the recognition rates before and after FS are shown in figure 6.16.

For all these different 2-class problems the features (both combined parameter and variance) number 0, 1, 4, 6 and 7 were removed. These features refer to the combined Eigenvectors $\varphi_{c,0}$, $\varphi_{c,1}$, $\varphi_{c,4}$, $\varphi_{c,6}$ and $\varphi_{c,7}$ which contain information about the largest variance. The recognition rates could be improved for all 5 age thresholds. This shows that some parts of largest variance are not relevant for the age evaluation whereas all the information about little variance is important.

4-Class Problem When removing 18 features, the recognition rate for the 4-class problem can be improved by 5% to 52.1%. The confusion matrix (see table 6.21) shows that age recognition is not an easy task. Class 0 (age group 0-19) is recognized fairly well (71.7%) whereas for all other classes at least 50% of the images are classified incorrectly.

	0	20	30	40
0	43	6	7	4
20	16	24	10	9
30	8	15	27	9
40	8	16	7	31 |//
Total error:		47.9%		

Table 6.21: Confusion matrix and error rate for the 4-class age evaluation

6.7.2.4 Selection of Images with Best Representation

Again, we select different subsets of images which can be represented well by the AAM. Here the recognition rates cannot be improved by selecting only images with a good TS value. Figure 6.17 shows an example of an image which is warped and synthesized. The image in the middle describes the error between these images. Although the TS value (0.97) is very high, the subjective AAM representation is not very good. The shape of glasses is visible in the synthesized image even though the person does not wear any glasses. This indicates that a good AAM representation is difficult for a database like the FG-NET Aging Database with many variations.

6.7.3 State-of-the-Art

Again, it is difficult to compare our age recognition rates to the results of other studies. Up to now, very little work was reported on age estimation from image data. Table 6.22 compares our 4-class result with the result of Wilhelm [120].

study	face databases	recognition rates
FEASy offline	FGNET-A	*52.1%*
AAM MLP 2005 [120]	NIFace2	~ 43%

Table 6.22: Comparison of our 4-class result with state-of-the-art age recognition

study	face databases	recognition rates
PCA-LDA 2005 [12]	FERET	91.5%
FEASy offline	FGNET-A	*76.0%*

Table 6.23: Comparison of our 2-class result with state-of-the-art age recognition

Wilhelm distinguishes between 5 classes also including an age group of 50 years and older on a dataset of 490 images. The database we used does not allow such a classification because the number of persons above 50 is very low (see figure 6.4). Taking into account that our result consists of 4 classes, the recognition rates are comparable.

Table 6.23 compares our 2-class result with the result made by Buchala [12]. The PCA-LDA recognition system, which performed worse for the gender recognition in comparison to our system (see subsection 6.3.3), is better for age recognition. However Buchala classifies the two age groups 20-39 and 50-60+, leaving a gap of 10 years. This simplifies the recognition task. Here our result is stated for the age groups of 0-29 and 30-69 which are more difficult to classify.

6.8 Comparison with NMF-AAMs

The NMF variant of the AAM algorithm was implemented by Michael Höchstetter [51] as part of the MMER Project. The same evaluation setups described in the sections before were used to obtain the results stated in table 6.24. For all NMF evaluations the SVM parameters were optimized. As for the PCA variant also a feature selection was performed for the age problems reducing the 2-class problem by 10 features and the 4-class problem by 18 features. For the 2-class age recognition we used the age threshold of 20.

problems	classes	PCA	NMF
gender recognition	2	94.6%	92.5%
facial expression	4	79.7%	74.8%
head pose	15	68.1%	48.8%
age	4	52.1%	46.1%
age	2	71.5%	68.7%

Table 6.24: Comparison of the PCA and NMF variant

The evaluations of the NMF variant were a first test and demonstration of the capabilities of this variant. The AAM parameters and the algorithm can be optimized to perform better (see [51]).
Nevertheless these results show that the NMF is a comparable alternative to the PCA as an AAM variant and shall be subject to future research.

6.9 Conclusion

In this chapter we presented the application and evaluation of different Active Appearance Model variants on several well established fields of face analysis, namely the recognition of gender, age, identity, head pose, and facial expression of a person in a digital image. We present a methodology for the selection of appropriate images for the generation of an Appearance Model. The following findings were drawn from our evaluations:

- In general the analysis of faces performs best, when the feature space of a statistical classification consists in the AAM coefficients (absolute values *and* values normalized to the dimension specific variance or Eigenvalues) plus the landmark positions in polar coordinates with the shape centroid as origin.

- There is a direct correlation between the quality of the re-synthesis measured by the texture similarity and the classification performance in pattern recognition tasks.

- In general the recognition rates based on AAM features can be increased by a preceding feature selection, since some tasks (e.g. age) require mainly AAM coefficients of dimensions with low variance and others with high variance (e.g. head pose).

- The best result for gender classification observed in our person disjunctive setup was 93.1%, while the facial expression recognition on the three classes smile, scream, and neutral showed 95.1% accuracy. The gender recognition on databases with persons from east-asia turned out to be more challenging than for caucasians.

- Unlike other analysis tasks, the AAMs could not keep up nor outperform the state-of-the-art algorithms in person identification.

- The head pose recognition performs best when an AAM is generated which covers only head poses to one side. Hence, this AAM is applied to the target image and its flipped version. Based on a better re-synthesis quality, the head pose is evaluated by the corresponding AAM features.

- The evaluations for age recognition ran on the challenging FG-Net Aging database. On this set recognition rates of 52.1% for 4 classes and 76.0% for 2 classes could be achieved.

- The Active Appearance Model based on Non-negative Matrix Factorization showed its functionality and applicability. The NMF allows for a more compact representation of the data variance and should therefore in further research efforts be applied to tasks with low global variance.

- The Nelder-Mead or Simplex algorithm was the most accurate optimization strategy under the investigated algorithms. However, in comparison to the offline predicted optimization a time factor of 10 lies in between and does therefore not meet real-time requirements any more.

According to our evaluations Active Appearance Model as one single approach proved to be equal or superior to state-of-the-art approaches especially developed for certain face analysis tasks. Solely person identification still seems especially challenging to AAMs. Since all the functionality is available in our software system FEASy, e.g. high level systems in human-machine interaction can exploit these analysis results provided that they are endued with the necessary computational power.

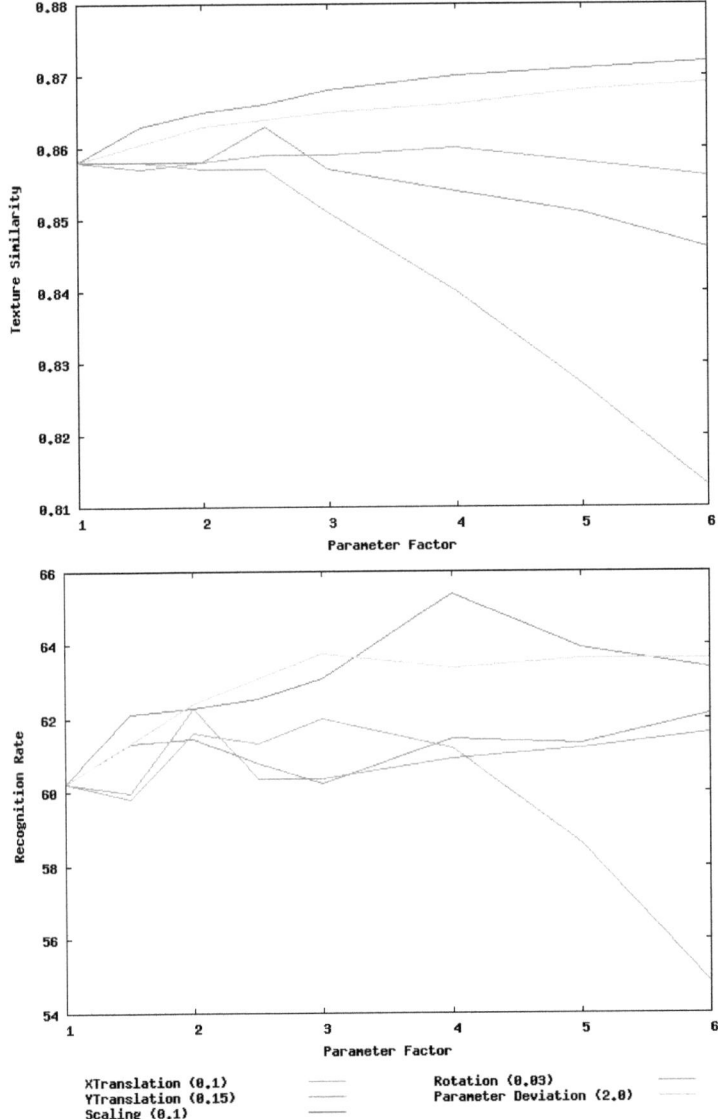

Figure 6.14: Mean TS values and recognition rates for different parameter variations (default values in brackets)

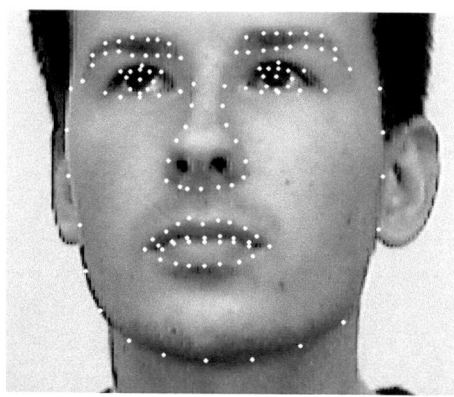

Figure 6.15: Example of the Störmer annotations with 132 landmarks

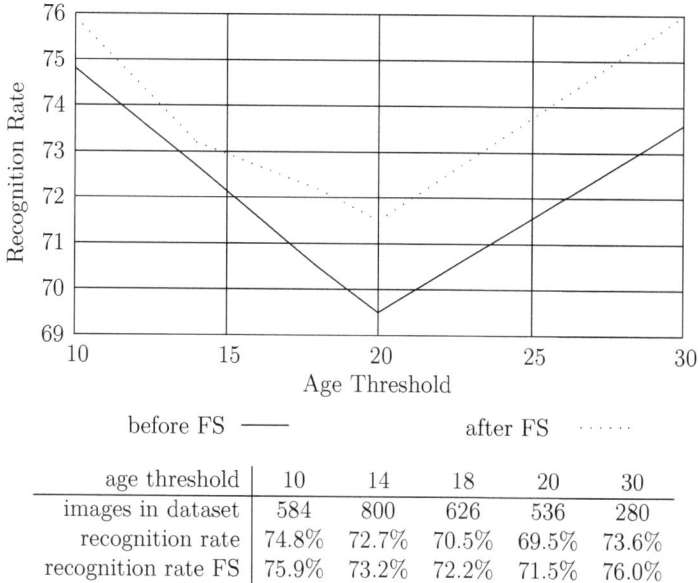

age threshold	10	14	18	20	30
images in dataset	584	800	626	536	280
recognition rate	74.8%	72.7%	70.5%	69.5%	73.6%
recognition rate FS	75.9%	73.2%	72.2%	71.5%	76.0%

Figure 6.16: Recognition rates for different age thresholds with and without Feature Selection

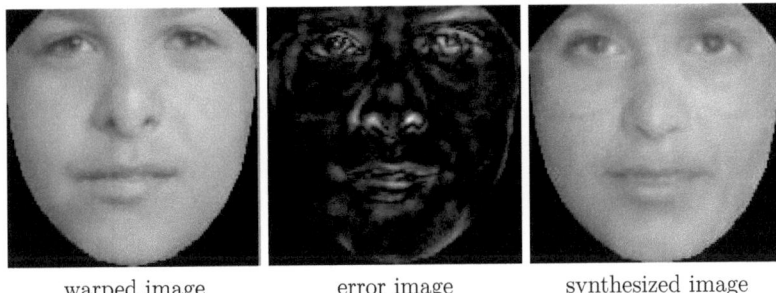

warped image   error image   synthesized image

Figure 6.17: Example of synthesis of an image with a Texture Similarity value of 0.97

Chapter 7

Summary

This dissertation introduced a software system for fully automatic face analysis called FEASy (sec. 1.2). The development of such system requires the workforce of a team of programmers and researchers. Therefore, the framework MMER_Lab was implemented which allows for a distributed development process on the one hand and a high-performance system execution at the other hand. These features were realized by a modular system structure and a multi-threading support for hardware optimized routines. On hardware architectures with dual-core CPUs an acceleration by 1/3 of serial signal processing systems could be observed (chapter 2).

Before it comes to the analysis of a face, its reliable localization is mandatory. Thus, a highly efficient and accurate method was developed and implemented which is based on a sampling of an image with windows of variable size. From each sample window visual Haar-like wavelet features are extracted. Thereby a Decision Stump as weak classifier operates on single features. These weak classifiers are combined by a Gentle AdaBoost which tries to reject windows without a face at early stages of a cascade. For initialization of our face analysis step a localization of the eyes runs on a narrowed area within the face region. It turned out that the addition of Gabor wavelet features and the replacement of the Decision Stump as weak classifier by an adaptive interval classifier leads to a localization at higher efficiency and smaller spatial deviation. Finally, according to our evaluations, 98.5% of the eyes in pictures with frontal human faces can be localized with less than 5 pixel Euclidean deviation from the actual eye center. The software implementation runs more than 5 times real-time on images of double VGA resolution. Therefore, the developed localization methodology provides a reliable basis of our fully automatic face analysis system (chapter 3).

Eventually, chapter 4 gives an introduction to the mathematical and implementation background of the standard Active Appearance Model (*abbrev.* 'AAM') algorithm which constitutes the basis for the face analysis. An Appearance Model statistically describes the variations in shape and texture of human faces derived from a careful selection of photographs. Depending on the specific application focus of the analysis these pictures should show different persons with different facial expressions and head poses in various lighting conditions. AAMs base on the assumption that the two-dimensional appearance of an object in a digital image is influenced by several widely independent sources. The target for the generation of an Appearance Model is to eliminate the variance induced by the camera view (translation, rotation, scale), illumination conditions, and the capturing device (brightness, intensity). Thus, the Appearance Model focuses on the independent modeling of the shape and texture in a first step and finally combines the two sources again to model the influences of shape variations on the texture. During the analysis of a human face within a video or a picture, the Appearance Model is used to re-synthesize this face as optimal as possible by adjustment of a set of scalar model coefficients.

Chapter 5 describes the variations, extensions, and optimizations which have been presented over the years as well as those which have been developed during the research for this work. The contributions address all stages of the Active Appearance Model algorithm, namely the model generation, the coefficient optimization, and the evaluation of the re-synthesis quality. Additionally, the aspects of an efficient implementation based on GPU involvement are covered. With the Non-negative Matrix Factorization a completely novel variant of Appearance Model generation is presented which can be subject of multiple further research activities. Standard AAMs suffer under the widely unrealistic precondition that the search space of the AAM coefficients is similar for all target objects under the whole range of variations in initializations, illuminations, and views. In order to overcome this constraints, several optimization strategies which operate without prediction schemes about the search space are investigated whereas the Nelder-Mead strategy turned out to be the best with respect to computational efficiency and quality of the re-synthesis.

Finally, the improved standard AAM algorithm plus the developed variants were applied to several well established fields of face analysis, namely the recognition of gender, age, identity, head pose, and facial expression of a person. Besides a methodology for the selection of appropriate images for the generation of Appearance Models, several further findings were drawn. These mainly regard the optimal algorithmic variant of AAMs, the preparation of AAM coefficients for classification, and the impact of the re-synthesis quality on the recognition performance. The evaluations showed that, apart from person identification, our single Active Appearance Model approach is equal or superior to other task specific state-of-the-art face analysis systems.

Since all the functionality is available in our software system FEASy, e.g. high level systems in human-machine interaction can exploit these analysis results provided that they are endued with the necessary computational power.

Further research may focus on strategies for the generation of Appearance Models which cover the entire range of ages, facial expressions, and ideally head poses of a single ethnic origin. Apart from image selection, an improved modeling of independent facial features will lead the way. Since Appearance Models with such variance are expected to show increasingly difficult search spaces, hybrid optimization methods which perform a rough minimum search followed by a fine-tuned local optimization will be required.

Part I

Appendix

Appendix A

Conventions

A.1 General Typesetting

For better readability we used specific standards for typesetting math symbols, functions or code fragments throughout this document.

A.1.1 Indexing

All indices for accessing sequences, ordered sets, vectors or matrices are zero based. For example let vector $\boldsymbol{v}$ with $|\boldsymbol{v}| = n$ be defined as

$$\boldsymbol{v} = \begin{pmatrix} v_0 & v_1 & \ldots & v_{n-1} \end{pmatrix}^T$$

A.1.2 Sets

Sets are formatted in capital calligraphic letters

$$\mathcal{H} = \mathcal{A} \cup \mathcal{B}$$

and members of a set are normally choosen so that letters match like

$$a \in \mathcal{A}, \quad b \in \mathcal{B}$$

A.1.3 Scalars

Scalar values are written in lower-case letters like

$$a = 5, \quad b = 3, \quad \alpha = \lambda \cdot \pi$$

A.1.4 Sequences

The index of sequences (which might be defined recursively) is always set in brackets

$$f_{[0]} = 0, \quad f_{[1]} = 1, \quad f_{[n]} = f_{[n-1]} + f_{[n-2]}$$

indicating that we are talking about the same f varying in discrete steps over time or over a number of iterations.

A.1.5 Functions

Functions are formatted either in (non-bold non-italic) lower-case or upper-case letters:
$$\mathrm{F}(x) = x^2, \quad \mathrm{f}(x) = 2x$$

A.1.6 Transformations

Transformations are typesetted just like functions:

$$\mathrm{T} : \boldsymbol{t}(\boldsymbol{x}) = \begin{pmatrix} a & b \\ c & d \end{pmatrix} \boldsymbol{x}$$

A transformation is applied (in this example to a vector $\boldsymbol{x}$) by
$$\mathrm{T}\{\boldsymbol{x}\}$$

Transformations can also be concatenated by the ∘ operator:
$$\mathrm{T}_2 \circ \mathrm{T}_1 \{\boldsymbol{x}\}$$

In this case, first transformation T_1 is applied to the vector $\boldsymbol{x}$, and then T_2 is applied to the result of the first transformation.

A.1.7 Vectors

Vectors are typesetted in bold lower-case letters
$$\boldsymbol{v} = \mu \boldsymbol{u}$$

where μ is a scalar value. If a distinctive vector is mentioned, its index is formatted in bold as well:
$$\boldsymbol{b_s} = \textit{some distinctive vector}$$

In the case we want to access a scalar element of a vector we write
$$b = \boldsymbol{b_s}_i, \quad c = \boldsymbol{c}_j$$

where b is the ith element of vector $\boldsymbol{b_s}$ and c is the jth element of vector $\boldsymbol{c}$. Please note that all vectors mentioned in this document are column vectors unless stated otherwise. Data consisting of several vectors is normally arranged column-wise in a matrix.

If we want to concatenate several vectors $\boldsymbol{v}_i$, $0 \leqslant i \leqslant n$ to form another vector $\boldsymbol{c}$, we write
$$\boldsymbol{c}^T = (\boldsymbol{v_0}^T \mid \ldots \mid \boldsymbol{v_n}^T)$$

A.1.8 Matrices

Matrices are formatted in bold capital letters

$$v = \alpha Ax + \beta b$$

where A is a matrix, x and b are vectors and α and β are scalars. If a distinctive matrix is mentioned, its index is formatted in bold as well:

$$Q_s = some\ distinctive\ matrix$$

In the case we want to access a scalar element of a matrix we write

$$q = Q_{s i,j}, \quad d = D_{k,l}$$

where q is the element in the ith row and jth column of matrix Q_s and d is the element in the kth row and lth column of matrix D. Sometimes we want to select a row or column vector of a matrix. In this case, the following notation is used:

$$u = A_{i,*}, \quad v = A_{*,j}$$

where u is the ith row vector of matrix A and v its jth column vector. If we want to point out clearly that the matrix M is made up of column vectors $\mu_i v_i$, $0 \leqslant i < n$, we write

$$M = (\mu_0 v_0 \mid \mu_1 v_1 \mid \ldots \mid \mu_{n-1} v_{n-1})$$

A.1.9 Code

Code fragments within the text are always written in typewriter font, for example to refer to the function foo() below. Listings are presented in a special environment like this:

```
1  foo() {
2      x=2
3      y=3
4      return x+y
5  }
```

A.1.10 Text Substitution

A variable string is usually enclosed into %...%. For example, %DIRECTORY% can name the system dependent folder where the application app can be found, like this:

```
%DIRECTORY%/bin/app
```

A.2 Symbols

For better syntactical understanding each mathematical symbol is only used once in a distinctive context throughout this document. See tables A.1 to A.5 for a list of all used symbols and their meaning.

Table A.1: Generic symbols

L	Number of landmarks
p	Number of all samples in the training set
d	Number of synthesized samples used for predictor matrix training
f	One-to-one mapping (bijective) function to associate images and their respective landmarks
E	Error function
T	Similarity transformation
e	Unit vector
$\boldsymbol{I}$	Identity matrix
$\boldsymbol{C}$	Covariance matrix

Table A.2: Shape model symbols

x	Annotation landmark (a tuple (x, y))
s	Shape vector in pixel space (containing interleaved coordinates)
$\mathcal{S}$	Set of original annotation point vectors
s'	Shape vector in normalized space (containing interleaved coordinates)
$\mathcal{S}'$	Set of aligned annotation point vectors
$\hat{s}$	Estimation for mean shape vector
$\overline{s}$	Mean shape vector in normalized space
S	Matrix of normalized shape vectors
C_s	Shape covariance matrix
μ_s	Number of used shape parameters
λ_s	Vector of shape model eigenvalues
Φ_s	Matrix of shape model eigenvectors (column-wise)
h_s	Shape model parameter vector

Table A.3: Texture model symbols

$\boldsymbol{p}$	Vector containing an original training image (in row-major-order)
$\mathcal{P}$	Set of training images
χ	Number of channels in an image
$\boldsymbol{e}$	Triangulation edge
$\mathcal{E}$	Set of triangulation edges
$\ominus$	Set of triangles
$\boldsymbol{t}$	Warped and masked texture vector (column vector)
$\mathcal{T}$	Set of warped and masked textures
c	Number of masked pixels
$\boldsymbol{t}'$	Normalized warped and masked texture vector (column vector)
$\mathcal{T}'$	Set of normalized warped and masked texture vectors
$\hat{\boldsymbol{t}}$	Estimation of mean texture vector
$\bar{\boldsymbol{t}}$	Mean texture vector
$\boldsymbol{T}$	Matrix of normalized warped and masked texture vectors
$\boldsymbol{C}_t$	Texture covariance matrix
μ_t	Number of used texture parameters
$\boldsymbol{\lambda}_t$	Vector of texture model eigenvalues
$\boldsymbol{\Phi}_t$	Matrix of texture model eigenvectors (column-wise)
$\boldsymbol{h}_t$	Texture model parameter vector

Table A.4: Combined model symbols

C_c	Combined model covariance matrix
μ_c	Number of used combined parameters
λ_c	Vector of combined model eigenvalues
Φ_c	Matrix of combined model eigenvectors (column-wise)
Φ_{cs}	Shape relevant sub-matrix of Φ_c
Φ_{ct}	Texture relevant sub-matrix of Φ_c
Q_s	Eigenvectors used for shape synthesis (column-wise)
Q_t	Eigenvectors used for texture synthesis (column-wise)
C	Matrix containing concatenated shape and texture parameter vectors for each sample
c	Column vector of C
h_c	Combined model parameter vector
k	Shape weight
K	Diagonal matrix containing the shape weight

Table A.5: Predictor matrix calculcation symbols

$\mathcal{V}$	A set of concatenated parameter vectors containing pose and texture parameters as well as the combined model parameter vector
$\boldsymbol{h}$	Concatenated parameter vector containing pose and texture parameters as well as the combined model parameter vector
v	Dimension of vector $\boldsymbol{h}$
$\boldsymbol{R}$	Predictor matrix
$\boldsymbol{J}$	Jacobian matrix
$\boldsymbol{r}$	Texture difference function
w	Gaussian kernel
s_r	Scale factor
θ	Rotation
t_x	Pose parameter for translation along the X axis
t_y	Pose parameter for translation along the Y axis
s	Pose parameter controlling scale
r	Pose parameter controlling rotation
b	Texture parameter for texture brightness
m	Texture parameter for texture intensity
α	Step width for gradient descent
β	Step width multiplier for gradient descent
ε	Termination criterion for search algorithm
T_p	Pose similarity transformation
T_g	Global similarity transformation

List of Figures

2.1	Schematic view of the framework concept	11
2.2	Training a SVM with AAM Parameter Data	14
3.1	1D Haar-Wavelet spectrum	19
3.2	2D-Haar-like Wavelets filters	19
3.3	Real part of a Gabor filter kernel	21
3.4	Family of Gabor-Wavelet filters	22
3.5	Structure of the threshold classifier	26
3.6	Exemplary threshold determination	27
3.7	Determination of intervals with the SPLIT strategy	28
3.8	Structure of cascades for object detection	29
3.9	Single stage of a cascade	29
3.10	Images of the FERET database and positive material for the eye localization	30
4.1	Variance in the appearance of human faces [1]	36
4.2	Annotation scheme for faces with 72 landmarks	36
4.3	(a) Source image with source shape s_i and triangle ψ - (b) position and shape of the target triangle ψ' in the mean shape $\overline{s}$	38
4.4	Effect of the first shape model components	41
4.5	Effect of the first two combined model components	43
4.6	Transformations during AAM coefficient optimization	46
4.7	Numerical Estimation of the Jacobian matrix	48
4.8	Schematic illustration of the AAM coefficient optimization algorithm	50
5.1	Graphical illustration of NMF	55
5.2	Illustration of theorem 5.2.2	57
5.3	Abstract dataset $\boldsymbol{D}_{circ}$	63
5.4	Face dataset $\boldsymbol{D}_{face}$	63
5.5	Base images of dataset $\boldsymbol{D}_{circ}$ after applying the PCA	63
5.6	Base images of dataset $\boldsymbol{D}_{circ}$ after applying the NMF	64
5.7	Base images of dataset $\boldsymbol{D}_{face}$ after applying the PCA	64
5.8	Base images of dataset $\boldsymbol{D}_{face}$ after applying the NMF	64
5.9	Reconstructed images (PCA)	65
5.10	Reconstructed images (NMF)	65
5.11	Face reconstruction	65

5.12	WarpingEngine2 rendering pipeline	78
5.13	The bounding box of an AAM shape s (destination vertices)	79
5.14	WarpingEngine2 scene setup: viewport, projection and viewing frustrum	80
5.15	WarpingEngine2 texture setup: virtual and physical texture	81
5.16	Incorrectly warped texture	83
5.17	Schematic illustration of CPU-based coefficient optimization	84
5.18	Schematic illustration of GPU-based coefficient optimization	84
5.19	Example of an AAM difference image	89
6.1	Example of different expressions from the AR Database	99
6.2	Example of different head poses from the NIFace1 Database	99
6.3	Example of different ages from the FG-NET Aging Database	100
6.4	histogram of the age variation	100
6.5	Example of the FG-NET Aging annotations with 68 landmarks	101
6.6	Texture Similarity histogram and evaluation results	103
6.7	Comparison of neutral (top) and anger (bottom) expression	105
6.8	Example of images (smile) taken 14 days apart	107
6.9	Different classes for the head pose evaluation	108
6.10	Example of an annotated image from the NIFace1 Database	109
6.11	Improvement of the AAM during AAM Search	110
6.12	AAMs for different head poses	112
6.13	Effect of the first 2 AAM principal components	113
6.14	Mean TS values and recognition rates for different parameter variations (default values in brackets)	123
6.15	Example of the Störmer annotations with 132 landmarks	124
6.16	Recognition rates for different age thresholds with and without Feature Selection	124
6.17	Example of synthesis of an image with a Texture Similarity value of 0.97	125

List of Tables

2.1	Evaluation results	15
3.1	Number of required features (overall #) and average number of computed features per sampled window on the test dataset (average #), as well as the number of Gabor features at application of both feature types (H+G) (G #)	31
3.2	Hit rate (HR) and average number of computed features per window on the test dataset (average #)	31
3.3	Average duration of the eye localization per face (in ms) regarding the used weak classifier	32
5.1	Distribution of convergence quality annotation	87
5.2	Confusion matrices for SVM classification of all quality measures and Texture Similarity only	91
6.1	Comparison of face databases	98
6.2	Cross-classification confusion matrices and error rates for gender evaluation; (a) non person-disjunctive and (b) person-disjunctive	102
6.3	Comparison of different optimization methods	102
6.4	Comparison of our result with state-of-the-art gender recognition rates	104
6.5	Cross-classification confusion matrices and error rates for facial expression evaluation; (a) non person-disjunctive and (b) person-disjunctive	105
6.6	Comparison of our result with state-of-the-art facial expression recognition rates	106
6.7	Results for the face identification	107
6.8	Comparison of our result with state-of-the-art face identification rates	108
6.9	TS values and recognition rates for different amounts of images for AAM training	110
6.10	Cross-classification confusion matrix and error rate for head pose evaluation with a 56 image AAM	111
6.11	Cross-classification confusion matrices and error rates for the 3-class head pose problem; (a) horizontal angles and (b) vertical angles	111

6.12 TS values and recognition rates for different combined percentages 112
6.13 TS values and recognition rates for different parameter combinations 113
6.14 Results for different 15-class problem systems 115
6.15 Evaluation results for images with best representation 115
6.16 Recognition rates with and without Head and Eye-Tracking Module 116
6.17 Comparison of our 5-class horizontal result with other horizontal head pose recognition rates 117
6.18 Comparison of our 3-class vertical result with other vertical head pose recognition rates 117
6.19 Recognition rates with and without manually improved annotations 118
6.20 TS values and recognition rates for different amounts of images for AAM training 118
6.21 Confusion matrix and error rate for the 4-class age evaluation . . 119
6.22 Comparison of our 4-class result with state-of-the-art age recognition 120
6.23 Comparison of our 2-class result with state-of-the-art age recognition 120
6.24 Comparison of the PCA and NMF variant 121

A.1 Generic symbols 134
A.2 Shape model symbols 135
A.3 Texture model symbols 136
A.4 Combined model symbols 137
A.5 Predictor matrix calculcation symbols 138

List of Listings

4.1	Jacobian matrix numerical estimation algorithm	48
5.1	Pseudocode of the NMF Algorithm	62

Bibliography

[1] FG-NET Aging Database. http://www-prima.inrialpes.fr/FGnet/.

[2] NIFace1 Database. http://tu-ilmenau.de/fakia/NIFace1.5255.0.html?&no_cache=1&sword_list[]=NIFace.

[3] ABATE, A., NAPPI, M., RICCIO, D., AND SABATINO, G. 2d and 3d face recognition: A survey. 1885–1906.

[4] ABELSON, HAROLD, AND WITH JULIE SUSSMAN, G. J. S. *Structure and Interpretation of Computer Programs (Second edition)*. The MIT Press, 1996, pp. –.

[5] AHLBERG, J. Using the active appearance algorithm for face and facial feature tracking. In *Proceedings of the IEEE Conference on Computer Vision* (2001).

[6] BAKER, S., AND MATTHEWS, I. Equivalence and efficiency of image alignment algorithms. In *Proceedings of the IEEE Conference on Computer Vision and Pattern Recognition* (2001).

[7] BASILI, P. Holistische Mimikerkennung mit Active Apperance Modellen. Bachelor thesis, Technische Universität München, Institute for Human-Machine Communication, 2006. Supervisor: Ronald Müller.

[8] BATUR, A., AND HAYES, M. A novel convergence scheme for active appearance models. In *Proceedings of the IEEE Conference on Computer Vision and Pattern Recognition* (2003).

[9] BATUR, A. U., AND HAYES, M. H. Adaptive active appearance models. *IEEE Transactions on Image Processing 14* (2005), 1707–1721.

[10] BLACKFORD, L. S., DEMMEL, J., DONGARRA, J., DUFF, I., HAMMARLING, S., HENRY, G., HEROUX, M., KAUFMAN, L., LUMSDAINE, A., PETITET, A., POZO, R., REMINGTON, K., AND WHALEY, R. C. An updated set of Basic Linear Algebra Subprograms (BLAS). *ACM Transactions on Mathematical Software 28*, 2 (June 2002), 135–151.

[11] BROWN, L., AND TIAN, Y. Comparative study of coarse head pose estimation. In *Motion02* (Hawthorne, NY 10532, 2002), pp. 125–130.

[12] BUCHALA, S., DAVEY, N., GALE, T. M., AND FRANK, R. J. Principal Component Analysis of gender, ethnicity, age, and identity of face images. *Proceedings of IEEE ICMI 2005* (2005).

[13] BUCIU, I., AND PITAS, I. Application of non-negative and local non negative matrix factorization to facial expression recognition. In *Proc. of the 17th Int. Conf. on Pattern Recognition* (Aug. 2004), vol. 1, pp. 288–291.

[14] BURGES, C. J. C. A tutorial on support vector machines for pattern recognition. *Data Mining and Knowledge Discovery 2*, 2 (1998), 121–167.

[15] CHARAMEL GMBH. mocito - application & content server for mobile services. https://www.mocito.com, 10 2007.

[16] CIGNONI, P., MONTANI, C., AND SCOPIGNO, R. DeWall: A fast divide & conquer Delaunay triangulation algorithm in E^d. *Computer-Aided Design 30*, 5 (1998), 333–341.

[17] CLEARY, J. G., AND TRIGG, L. E. K*: An instance- based learner using an entropic distance measure. In *Proceedings of the 12th International Conference on Machine learning* (1995), pp. 108–114.

[18] COLLOBERT, R., BENGIO, S., AND MARIÉTHOZ, J. Torch: a modular machine learning software library. Tech. Rep. 02-46, IDIAP, 2002.

[19] COOTES, T., EDWARDS, D., AND TAYLOR, C. Active appearance models. *IEEE Transactions on Pattern Analysis and Machine Intelligence 23* (2001), 681–685.

[20] COOTES, T., EDWARDS, G., AND TAYLOR, C. A comparative evaluation of active appearance model algorithms. In *Proceedings of the British Machine Vision Conference* (1998).

[21] COOTES, T., EDWARDS, G., TAYLOR, C., H.BURKHARDT, AND NEUMANN, B. Acitve appearance models. In *Proceedings of the European Conference on Comnputer Vision* (1998).

[22] COOTES, T., AND KITTIPANYA-NGAM, P. Comparing variations on the active appearance model algorithm. In *Proceedings of the British Machine Vision Conference* (2002).

[23] COOTES, T., AND TAYLOR, C. Constrained active appearance models. In *Proceedings of the IEEE Conference on Computer Vision* (2001).

[24] COOTES, T., AND TAYLOR, C. On representing edge structure for model matching. In *Proceedings of the IEEE Conference on Computer Vision and Pattern Recognition* (2001).

[25] COOTES, T., WALKER, K., AND TAYLOR, C. View-based active appearance modles. In *Proceedings of the IEEE Conference on Automatic Face and Gesture Recognition* (2000).

[26] COOTES, T. F., EDWARDS, G. J., AND TAYLOR, C. J. Active appearance models. In *IEEE Transactions on Pattern Analysis and Machine Intelligence* (June 2001), vol. 23, pp. 681–685.

[27] COOTES, T. F., AND TAYLOR, C. J. Statistical models of appearance for computer vision. Tech. rep., University of Manchester, UK, Mar. 2004.

[28] DAUGMAN, J. Two dimensional spectral analysis of cortical receptive field profile. In *Vision Research*, vol. 20. 1980, pp. 847–856.

[29] DELAUNAY, B. N. Sur la sphère vide. *Bulletin of Academy of Sciences of the USSR*, 6 (1934), 793–800.

[30] DEMPSTER, A., LAIRD, N., AND RUBIN, D. Maximum likelihood from incomplete data via the EM algorithm. *Journal of the Royal Statistical Society B 39*, 1 (1977), 1–38.

[31] DIDUCH, L. A Multi-Threading Framework for Distributed Development of Signal Processing Systems applied to Active Appearance Models. Diploma thesis, Technische Universität München, Institute for Human-Machine Communication, 2006.

[32] DIDUCH, L., GEISINGER, M. S., MÜLLER, R., NIKOLAUS, R. E., AND SCHLICHTER, M. The MMER Book. Tech. rep., Technische Universität München, Institute for Human-Machine Communication, Arcisstr. 21, D-80333 München, June 2006. http://www.mmer-systems.eu.

[33] DIDUCH, L., MÜLLER, R., AND RIGOLL, G. A framework for modular signal processing systems with high-performance requirements. In *Proceedings ICME* (July 2007).

[34] DONOHO, D., AND STODDEN, V. When does non-negative matrix factorization give a correct decomposition into parts. In *NIPS* (2003).

[35] EDWARDS, G., COOTES, T., AND TAYLOR, C. Advances in active appearance models. In *Proceedings of the IEEE Conference on Computer Vision* (1999).

[36] EDWARDS, G., COOTES, T., TAYLOR, C., BURKHARDT, H., AND NEUMANN, B. Face recognition using acitve appearance models. In *Proceedings of the European Conference on Computer Vision* (1998).

[37] EDWARDS, G., TAYLOR, C., AND COOTES, T. Interpreting face images using acitve appearance models. In *Proceedings IEEE Conference on Automatic Face and Gesture Recognition* (1998).

[38] EUROPEAN UNION, SIXTH FRAMEWORK PROGRAMME. Ami, augmented multipary interaction, integrated project. URL: www.amiproject.org, January 2004.

[39] FLEISS, J. L. *"Statistical methods for rates and proportions"*. 2nd ed. New York: John Wiley, 1981, 39-41, 2nd ed. New York: John Wiley, 1981.

[40] FOLEY, J. D., VAN DAM, A., FEINER, S. K., AND HUGHES, J. F. *Computer graphics: principles and practice (2nd ed.)*. Addison-Wesley Longman Publishing Co., Inc., Boston, MA, USA, 1990.

[41] FREUND, Y., AND SCHAPIRE, R. Experiments with a new boosting algorithm. In *International Conference on Machine Learning* (1996), pp. 148–156.

[42] FRIEDMAN, J., HASTIE, T., AND TIBSHIRANI, R. Additive logistic regression: A statistical view of boosting.

[43] GAO, W., CAO, B., SHAN, S., ZHANG, X., AND ZHOU, D. The CAS-PEAL large-scale chinese face database and baseline evaluations. Tech. rep., JDL, 2004.

[44] GAO, Y., LEUNG, M., HUI, S., AND TANANDA, M. Facial expression recognition from line-based caricatures. *SMC-A 33*, 3 (May 2003), 407–412.

[45] GEISINGER, M. S. MMER_Lab: A Framework for modular Assembly of High-performance Signal Processing Systems. Diploma thesis, Technische Universität München, Informatik VI: Robotics and Embedded Systems, Institute for Human-Machine Communication, Oct. 2007.

[46] GEISINGER, M. S., AND SCHLICHTER, M. Convergence optimization of a system for face analysis. Tech. rep., Technische Universität München, Institute for Human-Machine Communication, Arcisstr. 21, D-80333 München, 2006.

[47] GOURIER, N., HALL, D., AND CROWLEY, J. L. Estimating Face Orientation from Robust Detection of Salient Facial Features. In *Proceedings of Pointing 2004, ICPR, International Workshop on Visual Observation of Deictic Gestures* (2004).

[48] GROSS, R., SHI, J., AND COHN, J. Quo vadis face recognition? - the current state of the art in face recognition. Tech. rep., Robotics Institute, Carnegie Mellon University, Pittsburgh, PA, USA, 2001.

[49] HAMMERSTONE, R., CRAIGHEAD, M., AND AKELEY, K. ARB_vertex_buffer_object OpenGL extension specification. SGI OpenGL Extension Registry, Jan. 2003. http://oss.sgi.com/projects/ogl-sample/registry/ARB/vertex_buffer_object.txt.

BIBLIOGRAPHY 149

[50] HIPP, D. R., AND CRUSE, M. Pttcl: Using tcl with ptthreads. In *Proceedings of the 5th annual Tcl/Tk Workshop* (1997).

[51] HÖCHSTETTER, M. Extension of the Active Appearance Model Approach by Non-negative Matrix Factorization. Diploma thesis, Technische Universit{at M}unchen, Institute for Human-Machine Communication, 2007.

[52] HOU, X., LI, S., ZHANG, H., AND CHENG, Q. Direct appearance modles. In *Proceedings of the IEEE Conference on Computer Vision and Pattern Recognition* (2001).

[53] HOYER, P. O. Non-negative matrix factorization with sparseness constraints. *CoRR cs.LG/0408058* (Aug. 2004).

[54] HOYER, P. O. Non-negative matrix factorization with sparseness constraints, Aug 2004.

[55] INTEL CORPORATION. Intel(R) software insight - multi-core capability. www.intel.com/cd/software/main/asmo-na/eng/285893.htm, 2005.

[56] JARRE, F., AND STOER, J. *Optimierung.* Springer, 2003.

[57] JONES, J., AND PALMER, L. An evaluation of the two-dimensional gabor filter model of simple receptive fields in cat striate cortex. In *Journal of Neurophysiology*, vol. 58. 1987, pp. 1233–1258.

[58] KANWISHER, N., MCDERMOTT, J., AND CHUN, M. M. The fusiform face area: A module in human extrastriate cortex specialized for face perception. In *Journal of Neuroscience* (June 1997), S. for Neuroscience, Ed., vol. 17, pp. 4302–4311.

[59] KEMBHAVI, A. Gender recognition. Tech. rep., Institute for Advanced Computer Studies, University of Maryland, USA, 2005.

[60] KEMPF, W. E. The boost threads library. *C/C++ Users Journal* (2002).

[61] KILGARD, M. J. NV_texture_rectangle OpenGL extension specification. SGI OpenGL Extension Registry, Mar. 2004. http://oss.sgi.com/projects/ogl-sample/registry/NV/texture_rectangle.txt.

[62] KOHAVI, R. A study of cross-validation and bootstrap for accuracy estimation and model selection. In *Proceedings of the Fourteenth International Joint Conference on Artificial Intelligence* (1995), vol. 2, pp. 1137–1143.

[63] KRIEGEL, S. The Application of Active Appearance Models to Comprehensive Face Analysis. Bachelor thesis, Technische Universität München, Institute for Human-Machine Communication, Apr. 2007.

[64] LEE, D. D., AND SEUNG, H. S. Learning the parts of objects by non-negative matrix factorization. *Nature 401*, 6755 (October 1999), 788–791.

[65] LEE, D. D., AND SEUNG, H. S. Algorithms for non-negative matrix factorization. In *NIPS* (2000), pp. 556–562.

[66] LEE, T. Image representation using 2d gabor wavelets. *IEEE Transactions on Pattern Analysis and Machine Intelligence 18*, 10 (October 1996), 959–971.

[67] LENGYEL, E. *OpenGL Extensions Guide*, first ed. Charles River Media, Graphics Series, July 2003.

[68] LI, S., CHENG, Y., ZHANG, H., AND CHENG, Q. Multi-view face alignment using direct appearance models. In *Proceedings of the IEEE Conference on Automatic Face and Gesture Recognition* (2002).

[69] LIU, C., AND WECHSLER, H. Independent component analysis of gabor features for face recognition. In *Transactions on Neural Networks*, vol. 14. IEEE, 2003, pp. 919–928.

[70] MA, B., YANG, F., GAO, W., AND ZHANG, B. The application of extended geodesic distance in head poses estimation. In *ICB* (2006), pp. 192–198.

[71] MARTINEZ, A., AND BENAVENTE, R. *The AR Face Database*, vol. 24 of *CVC Technical Report*. June 1998.

[72] MARTINEZ, A. M. Recognizing expression variant faces from a single sample image per class. *IEEE Int. Conf. on Computer Vision and Pattern Recognition (CVPR)* (2003).

[73] MARTINEZ, A. M., AND BENAVENTE, R. The AR face database. Tech. Rep. 24, CVC, June 1998.

[74] MCGLAUN, G., LANG, M., AND RIGOLL, G. Development of a generic multimodal framework for handling error patterns during HMI. In *Proc. of SCI* (2004), vol. I.

[75] MEHRABIAN, A. Communication without words. *Psychology Today 2*, 4 (1968), 53–56.

[76] MENACHER, A. Anwendung des Viola-Jones Algorithmus zur Detektion von Sicherungsblicken im Fahrzeug. Diploma thesis, Technische Universität München, Institute for Human-Machine Communication, Mar. 2006. Supervisor: Ronald Müller.

[77] MICHEL, M., STANFORD, V., AND GALIBERT, O. Network transfer of control data: An application of the NIST smart data flow. *Journal of Systemics, Cybernetics and Informatics 2*, 6 (2005).

BIBLIOGRAPHY

[78] MITCHELL, S., LELIEVELDT, B., GEEST, R., BOSCH, J., REIBER, J., AND SONKA, M. Multistage hybrid active appearance model matching: Segmentation of left and right ventricles in cardiac mr images. *IEEE Transactions on Medical Imaging 20* (2001), 415–423.

[79] MITCHELL, S., LELIEVELDT, B., GEEST, R., BOSCH, J., REIBER, J., AND SONKA, M. 3-d active appearance models: Segmentation of cardiac mr and ultrasound images. *IEEE Transactions on Medical Imaging 21* (2002), 1167–1178.

[80] MÜLLER, R., NIKOLAUS, R., GEISINGER, M., DIDUCH, L., SCHLICHTER, M., AND RIGOLL, G. A modular multi-threading system for automatic face synthesis and analysis. In *MLMI* (Washington D.C., USA, 2006), V. Stanford, Ed., Demonstration.

[81] MÜLLER, R., NIKOLAUS, R., GEISINGER, M., DIDUCH, L., SCHLICHTER, M., AND RIGOLL, G. A system for facial expression analysis. In *Wainhouse Research European Forum 2006* (Berlin, Germany, 2006), C. Perey, Ed., Demonstration, EU FP6 Integrated Project Augmented Multipary Interaction AMI.

[82] MOGHADDAM, B., AND YANG, M.-H. Learning gender with support faces. In *IEEE Transactions on Pattern Analysis and Machine Intelligence* (Grenoble, France, 2005), vol. 24, no.5.

[83] NELDER, J., AND MEAD, R. A simplex method for function minimization. *The Computer Journal 7* (1965), 308–313.

[84] NIKOLAUS, R. E. Design and Development of a GPU-accelerated Face Analysis System. Diploma thesis, Technische Universität München, Informatik VI: Robotics and Embedded Systems, Institute for Human-Machine Communication, Nov. 2007.

[85] NORDSTRØM, M. M., LARSEN, M., SIERAKOWSKI, J., AND STEGMANN, M. B. The IMM face database - an annotated dataset of 240 face images. Tech. rep., Informatics and Mathematical Modelling, Technical University of Denmark, DTU, Richard Petersens Plads, Building 321, DK-2800 Kgs. Lyngby, May 2004.

[86] NVIDIA CORPORATION. *NVIDIA CUDA - Compute Unified Device Architecture Programming Guide.* NVIDIA Corporation, 2701 San Tomas Expressway, Santa Clara, CA 95050, June 2007. http://developer.nvidia.com/object/cuda.html.

[87] NVIDIA CORPORATION. *NVIDIA CUDA - CUBLAS Library.* NVIDIA Corporation, 2701 San Tomas Expressway, Santa Clara, CA 95050, June 2007. http://developer.nvidia.com/object/cuda.html.

[88] OUSTERHOUT, J. K. *Tcl und Tk. Entwicklung grafischer Benutzerschnittstellen fur das X Window System.* Addision Wesley Proffesional Computing Series, 1997.

[89] PANTIC, M., VALSTAR, M. F., RADEMAKER, R., AND MAAT, L. Web-based database for facial expression analysis. In *Proc. IEEE Int'l Conf. on Multimedia and Expo (ICME'05)* (Amsterdam, The Netherlands, July 2005), pp. 317–321.

[90] PASCUAL-MONTANO, A., CARAZO, J., KOCHI, K., LEHMANN, D., AND PASCUAL-MARQUI, R. D. Nonsmooth nonnegative matrix factorization. *IEEE Transactions on Pattern Analysis and Machine Intelligence 28*, 3 (2006), 403–415.

[91] PASCUAL-MONTANO, A., CARAZO, J., KOCHI, K., LEHMANN, D., AND PASCUAL-MARQUI, R. D. Nonsmooth nonnegative matrix factorization (nsNMF). *IEEE Transactions on Pattern Analysis and Machine Intelligence 28*, 3 (2006), 403–415.

[92] PETEREIT, L. Prinzipien der bildverarbeitung im visuellen system des menschen. Proseminar Computer Vision Universität Ulm, 2004.

[93] PHILLIPS, P., MOON, H., RIZVI, S., AND RAUSS, P. The feret evaluation methodology for face-recognition algorithms. *PAMI 22*, 10 (October 2000), 1090–1104.

[94] QUINLAN, R. *C4.5: Programs for Machine Learning.* Morgan Kaufmann Publishers, San Mateo, CA, 1993.

[95] REIDSMA, D., POPPE, R., AND RIENKS, R. Parlevision, rapid development of computer vision applications, 2005.

[96] SAATCI, Y., AND TOWN, C. Cascaded classification of gender and facial expression using active appearance models. In *FGR06* (2006), pp. 393–400.

[97] SAMARIA, F., AND HARTER, A. Parameterisation of a stochastic model for human face identification. In *WACV94* (1994), pp. 138–142.

[98] SANDMEL, J., AND KILGARD, M. J. ARB_texture_non_power_of_two OpenGL extension specification. SGI OpenGL Extension Registry, May 2004. http://oss.sgi.com/projects/ogl-sample/registry/ARB/texture_non_power_of_two.txt.

[99] SCHAPIRE, R., AND SINGER, Y. Improved boosting algorithms using confidence-rated predictions. In *Proceedings of the Eleventh Annual Conference on Computational Learning Theory* (1998).

[100] SCHMIDT, D., STAL, M., ROHNERT, H., AND BUSHMANN, F. *Pattern Oriented Software Architecture. Patterns for Concurrent and Networked Objects. Volume 2.* John Wiley and Sons, 2000, pp. 343–355.

BIBLIOGRAPHY

[101] SCHULLER, B. *Automatische Emotionserkennung aus sprachlicher und manueller Interaktion.* PhD thesis, Technische Universität München, 2006.

[102] SEGAL, M., AND AKELEY, K. *The OpenGL(R) Graphics System: A Specification – (Version 2.0 - October 22, 2004),* Oct. 2004. http://www.opengl.org.

[103] SHREINER, D., WOO, M., NEIDER, J., AND DAVIS, T. *OpenGL(R) Programming Guide: The Official Guide to Learning OpenGL(R), Version 2,* fifth ed. Addison-Wesley Professional, August 2005.

[104] SIM, T., BAKER, S., AND BSAT, M. The CMU pose, illumination, and expression (pie) database of human faces. In *Proc.Int'l Conf. on Automatic Face and Gesture Recognition* (2002).

[105] STEGMANN, M. Object tracking using active appearance models. In *Proceedings of the Danish Conference on Pattern Recognition and Image Analysis* (2001).

[106] STEGMANN, M., AND LARSEN, R. Multi-band modeling of appearance. *Image and Visual Computing 21* (2003), 61–67.

[107] STEGMANNN, M., FISKER, R., AND ERSBLL, B. Extending and applying active appearance modles for automated, high precision sementation in different image modalities. In *Proceedings of the Scandinavian Conference on Image Analysis* (2001).

[108] STIEFELHAGEN, R. Estimating head pose with neural networks - results on the Pointing04 ICPR workshop evaluation data. In *Pointing 04 ICPR Workshop* (Cambridge, UK, August 2004).

[109] STÖRMER, A. Untersuchung der möglichkeiten zur beschreibung von gesichtern mit einer minimalen anzahl von appearance-parametern. Diploma thesis, Technische Universität Ilmenau, Fakultät für Informatik und Automatisierung, Mar. 2004.

[110] STÖRMER, A., AND STADERMANN, J. Constructing synthetic faces using Active Appearance Models and evaluating the similarity to the original image data. *Fortschrittberichte VDI 10,* 743 (2006), 47–56.

[111] TIA, Y., KANADE, T., AND COHN, J. Evaluation of gabor-wavelet-based facial action unit recognition in image sequences of increasing complexity. In *Proceedings of the Fifth IEEE International Conference on Automatic Face and Gesture Recognition* (2002).

[112] TIVIVE, F. H., AND BOUZERDOUM, A. A gender recognition system using shunting convolutional neural networks. *2006 International Joint Conference on Neural Networks* (2006).

[113] TURK, M., AND PENTLAND, A. Eigenfaces for recognition. *Journal of Cognitive Neuroscience 3*, 1 (Mar. 1991), 71–86.

[114] VAPNIK, V. N. *The nature of statistical learning theory*, 2nd ed. Springer, 1995.

[115] VIOLA, P., AND JONES, M. L. Robust real-time object detection. In *Cambridge Research Laboratory, Technical Report Series* (2001).

[116] VUGRIN, K. E. *On the Effects of Noise on Parameter Identification Optimization Problems*. PhD thesis, Faculty of the Virginia Polytechnic Institute and State University, Blacksburg, Virginia, April 2005.

[117] WANG, Y., JIA, Y., HU, C., AND TURK, M. Fisher non-negative matrix factorization for learning local features. In *Asian Conference on Computer Vision* (January 2004).

[118] WANG, Y., JIA, Y., HU, C., AND TURK, M. Non-negative matrix factorization framework for face recognition. *International Journal of Pattern Recognition and Artificial Intelligence Vol. 19, No. 4* (2005), 1–17.

[119] WHALEY, R. C., AND PETITET, A. Minimizing development and maintenance costs in supporting persistently optimized BLAS. *Software: Practice and Experience 35*, 2 (Feb. 2005), 101–121. http://www.cs.utsa.edu/~whaley/papers/spercw04.ps.

[120] WILHELM, T., BÖHME, H.-J., AND GROSS, H.-M. Classification of face images for gender, age, facial expression, and identity. *Proc. Int. Conf. on Artificial Neural Networks (ICANN'05), Warsaw* (2005).

[121] WITTEN, I. H., AND FRANK, E. *Data Mining: Practical machine learning tools and techniques*, second ed. Morgan Kaufmann, San Francisco, 2005.

[122] WU, B., AI, H., HANG, C., AND LAO, S. Fast rotation invariant multiview face detection based on real adaboost.

[123] WU, C. On the convergence properties of the EM algorithm. *The Annals of Statistics 11*, 1 (1983), 95–103.

[124] WÜSTNER, L. Visuelle Lokalisation und Klassifikation mittels AdaBoost-Varianten auf Basis von Haar- und Gabor-Wavelets. Diploma thesis, Technische Universität München, Institute for Human-Machine Communication, Mar. 2006. Supervisor: Ronald Müller.

[125] XIE, X., AND LAM, K.-M. An efficient method for face recognition under varying illumination. *IEEE International Symposium on Circuits and Systems vol. 4* (2005), 3841–3844.

[126] YAN, S., LIU, C., LI, S., ZHANG, H., SHUM, H., AND CHENG, Q. Face alignment using texture-constrained active shape models. *Image and Visual Computing 21* (2003), 69–75.

[127] ZHI, Y., AND MING, G. A SOM-wavelet networks for face identification. In *ICME 2005* (Amsterdam, July 2005).

Die VDM Verlagsservicegesellschaft sucht für wissenschaftliche Verlage abgeschlossene und herausragende

Dissertationen, Habilitationen, Diplomarbeiten, Master Theses, Magisterarbeiten usw.

für die kostenlose Publikation als Fachbuch.

Sie verfügen über eine Arbeit, die hohen inhaltlichen und formalen Ansprüchen genügt, und haben Interesse an einer honorarvergüteten Publikation?

Dann senden Sie bitte erste Informationen über sich und Ihre Arbeit per Email an *info@vdm-vsg.de*.

Sie erhalten kurzfristig unser Feedback!

VDM Verlagsservicegesellschaft mbH
Dudweiler Landstr. 99   Telefon +49 681 3720 174
D - 66123 Saarbrücken   Fax   +49 681 3720 1749
www.vdm-vsg.de

Die VDM Verlagsservicegesellschaft mbH vertritt

Printed by Books on Demand GmbH, Norderstedt / Germany